Food and Human Responses

Kodoth Prabhakaran Nair

Food and Human Responses

A Holistic View

 Springer

Kodoth Prabhakaran Nair
Villament G3
C/o Dr. Mavila Pankajakshy
Malaparamba, Kozhikode, Kerala, India

ISBN 978-3-030-35436-7   ISBN 978-3-030-35437-4   (eBook)
https://doi.org/10.1007/978-3-030-35437-4

This Springer imprint is published by the registered company Springer Nature Switzerland AG.
The registered company address is: Gewerbestrasse 11, 6330 Cham, Switzerland

I dedicate this book, which was compiled under very trying circumstances, to the memory of my parents,
my late father, Kuniyeri Pookkalam Kannan Nair, an illustrious police officer who served the British Police and who was decorated with King George V Medal for Bravery and Honesty,
and my late mother, Kodoth Padinhareveetil Narayani Amma, daughter of the aristocratic Kodoth family of North Malabar,
Kerala State, India, both of whom left me an orphan at a young age but whose boundless love and blessings made me what I am today.
My wife, Dr. Pankajam Nair, and our children, Dr. Kannan Mavila, Engineer

Sreedevi Mavila, and Engineer Arvind, have been a constant source of encouragement to me during the finalization of the contents and writing of this book.
Sasha, our daughter Sreedevi's Cocker-Spaniel, is a source of great joy, to me, whenever my wife and I visit our daughter in Hyderabad, India. I dedicate this book to Sasha, as well.

Note of Appreciation of the Book: Food and Human Responses: A Holistic View

Dr. K. V. Raman Ph.D. (Wisconsin)

I was immensely pleased when I learnt that Professor Kodoth Prabhakaran Nair, nominated for the 2017 Volvo Environment Prize, chose to compile a manuscript for a book titled *Food and Human Responses: A Holistic View*. Professor Nair has spent over three decades working in Europe, Africa, and Asia. A senior Humboldt Fellow, he was the first and only Asian scientist named to the prestigious National Chair of the Science Foundation, The Royal Society, Belgium, following the development and presentation of a revolutionary soil management concept, now globally known as "The Nutrient Buffer Power Concept," in the 6th "International Colloquium for the Optimization of Plant Nutrition," held in Montpellier, France, in 1984. He headed the Departments of Agriculture, Soil Sciences, and Basic Sciences at the University Center, Republic of Cameroon, from 1983 to 1991 and was later invited as senior professor to the Republic of South Africa to help build a Faculty of Agriculture at the prestigious University of Fort Hare, Alice *(alma mater of late Mr. Nelson Mandela, from where Mr. Mandela spearheaded the anti-apartheid struggle, when he was a law student in the University)*.

The book has 14 chapters covering a range of topics on important perspectives on food and health. Tracing the significance of food in many cultures, some of the chapters such as conceptual models in food-behavior relationships, nutrition-behavior interface and the way forward in research in this area, the brain-behavior link, the role of cholesterol in human health, examples of flagrant violations of the ethics code in dealing with food, the protein-energy malnutrition(PEM) among the poor of Africa, and the consequent diseases like kwashiorkor and marasmus, dietary supplements and their commercial exploitation, and finally the unethical treatment of food as a corporate target, are, indeed, illuminating. Professor Nair must be congratulated for his painstaking research, spanning more than 5 years, tenacity of purpose, and admirable scholarship in understanding these problems and as he rightly points out in the book the paradox of gluttony of the rich and the pity of hunger of the poor, coexisting among the rich-poor divide, even in countries like India, which is supposedly "self-sufficient" in food production, thanks to the highly soil extractive "green revolution," though with its disastrous environmental fallout, especially on soil, which Professor Nair is attempting to mitigate through his revolutionary soil management technique. "The Nutrient Buffer Power Concept," for which he recently received the **Swadeshi Sastra Puraskar**, of the Swadeshi Science Movement of the Government of India which is a forum to highlight and recognize admirable initiatives in science for the betterment of the country.

The book brings to light a hitherto superficially and inadequately discussed theme of food-behavior interface. While recommending the book to a global readership, I would particularly commend it to those scientists, students, and policymakers who deal with food and nutrition issues of relevance to the community at large. I must congratulate Prof. Nair for his originality of thinking, tenacious research, comprehensive understanding, and, above all, commendable simplicity in explaining even the most complex concepts in free-flowing style. I wish the book much success.

Member, Agricultural Scientists   (K. V. Raman)
Recruitment Board, New Delhi (Retd.)   36, Fourth Seaward Road,
Director, National Academy of Agricultural   Chennai, India
Research Management, Hyderabad   Chennai-600041 India
Dean and Professor, Pantnagar and   Email: kvraman100@yahoo.com
Andhra Pradesh Agricultural Universities

Donald L. Sparks

I am pleased to learn that Dr. Nair, an internationally recognized agricultural scientist, has written a book titled *Food and Human Responses: A Holistic View*. The book has 14 chapters, covering a wide spectrum of subjects, ranging from food production changes to the perils of under- and overnutrition. The book will be useful to scientists, students, and professionals engaged in an area of nutrition and its behavioral impact.

S. Hallock du Pont Chair in Soil
and Environmental Chemistry,
Francis Alison Professor, and
Director, Delaware Environmental Institute
S. Hallock du Pont Chair,
Francis Alison Professor, Professor
of Plant and Soil Sciences,
Chemistry and Biochemistry, Civil
and Environmental Engineering,
and Marine Science and Policy,
and Director, Delaware Environmental Institute
(DENIN)

Donald L. Sparks
221 Academy Street
Room 250A
Newark, DE 19716
Phone: 302/831-3436
Email: dlsparks@udel.edu

Dr. Krishnan Varikara

I am extremely delighted to learn that Professor Kodoth Padinhareveetil Prabhakaran Nair has compiled a book titled *Food and Human Responses: A Holistic View*.

Professor Nair is an internationally acclaimed agricultural scientist who set out to write on a theme that touches on crucial aspects of human life vis-à-vis the food a person ingests.

Human nutrition has been an evolutionary subject. It took millions of years for human beings to evolve nutritionally as well as observe its effects on physical characteristics, constitution, as well as mental and intellectual development.

The discovery of farming has contributed substantially toward the availability of food round the year. While life spreads out into colder areas, there was more dependence on dairy and meat, as farming throughout the year was not possible in the extreme weather conditions of the temperate regions. Similarly, the invention of cooking and knowledge of preservation and storage of food also modified our diet pattern considerably. We could get the benefit of large amounts of macro- and micronutrients in a small meal with cooked food. This knowledge has taken us forward in keeping us in good physical, mental, and intellectual health. Unfortunately, this has gone to the extent of tipping the fine balance between health and overnutrition and its consequential impacts.

The development of science and research has thrown light on microspore of the human body and gut determining its role on the evolution of diseases, autoimmunity, and genetic mutations behind diseases. The altered microspore of the gut based on diet has started revealing the myth behind many lifestyle, neoplastic, autoimmune, behavioral, and psychiatric diseases.

What amazes me about the book is the comprehensive information it contains, touching almost every aspect of nutrition which will give the reader an idea on this evolutionary subject to make rational choices on one's eating to live healthy.

The book contains 14 chapters which touch upon important perspectives on food and health; conceptual models in food-behavior relationships; nutrition-behavior interface; brain-behavior link; the role of cholesterol in human health; malnutrition; the contours of protein-energy malnutrition (PEM) and its fallout on academic performance of school children; the curse of food insecurity for the poor and the needless gluttony of the rich; the role of mineral nutrition in good health and the role of external stimulants, including alcohol on human behavior; and, finally, food as a corporate target. These are some of the very illustrative and illuminating chapters of the book.

On a personal note, I might add that Professor Nair and I have had many interactions on academic discussions on the interplay of nutrition and illness. I admire his scholarship and tenacity of purpose in attempting to chronicle some of the unknown links of food-human response interface.

As a practicing physician and academic who has spent more than a quarter century in understanding diseases, I feel that the book brings to the reader a new dimension in food-behavior interface. I would wholeheartedly recommend this book to a vast readership of global public with diverse eating culture. Besides this, academicians, researchers, and students in the field of nutrition and related faculty will be able to initiate new research and development to take this hypothesis to the extent of making the right rational policies in public health in the current century where the health-care burden is threatened by obesity and lifestyle and metabolic diseases.

Professor Nair has a checkered academic career spanning more than five decades in Europe, Africa, and Asia. A senior Alexander von Humboldt Fellow, he was the first and only Asian scientist to hold the prestigious National Chair of the Science Foundation, The Royal Society, Belgium.

Professor Nair has published extensively, and I had the opportunity to read some of his books including the one titled "Issues in National and International Agriculture" which was launched by India's great president, late Dr. Abdul Kalam, who was the chair of the National Science Forum.

I congratulate and wish Professor Nair all success in the launch of this book, in taking the message it contains for the betterment of physical, mental, and spiritual health globally and for the elimination of poverty, bringing peace and harmony over the world.

MBBS, MD (Gen Med),   Dr. Krishnan Varikara
MDC (Diabetes Care), FRACP
Consultant Physician, Lyell McEwin Hospital,
Elizabeth Vale, SA
Senior Lecturer and Conjoint Faculty Medicine
University of Adelaide, SA,
and University of Newcastle, NSW

A Word of Appreciation

I am very delighted to gratefully acknowledge the very enthusiastic support of Ms. Margaret Deignan, Senior Publisher, Springer, on this book project, and the dedicated and diligent support of Ms. Malini Arumugam, as Project Manager, in the book production. It has been a great pleasure working with Ms. Malini Arumugam.

Preface

Henry Kissinger, former secretary of the USA during the administration of late President Nixon, used to say "Control oil and you control nations; control food and you control people." Food becomes so very important to human life. That is the reason why I chose it as a subject for this book and wrote about it not from a mundane angle but with a totally different perspective as to how it affects human responses, in terms of the biochemistry of nutrition, linked to the central nervous system. Being an agricultural scientist, it helped.

I have always wondered, observing people, during the last more than five decades of my life, why some are, almost always, tranquil and cheerful in their demeanor and interactions with others around them, while quite a few are always "on edge," bordering on anger and intemperance, even during inconsequential verbal interactions with others, where there is the tendency to "hurt" others, not physically but verbally. Does it have to do anything with the food they consume? To put it more bluntly, does our "behavior" depend on the food we ingest? If I should give this thought a "Hindu twist," it has all to do with our past *karma* accumulated over millennia during our previous births, which we are not aware of in the present birth, where all the *satwic* (pure, unpolluted thoughts and actions), *rajasic* (greedy, violent, intemperate thoughts and actions), and *tamasic* (lethargic, indifferent thoughts and actions) karmas shape our current lives. Dr. Brian Weiss, the distinguished American medical psychologist, has beautifully illustrated this in his best-selling book *Many Lives, Many Masters*. But where does food come in this whole picture? Being an international agricultural scientist, my job concerns the production of food, ensuring sustainable use of the land on which that food is produced. So, the thought struck me why not I research the area where food and human behavior are interconnected. This meant that I must explore all the factors which connect both, involving biochemical pathways of food in the human body, the role of the central nervous system, food consumption-related diseases such as anorexia nervosa (fear of food as an agent to excessive weight gain and so deliberate fasting) and bulimia nervosa (gluttony and consequent purging), the role of sugar in human health, alcoholism, and, finally, overweight and obesity – the bane of modern times. To understand better the entire gamut of food-behavior linkage, one has also to unravel the workings of the mind-brain link. I have made an honest attempt in understanding this aspect, which I hope adds value to the already existing fund of knowledge in this area. Much of today's bodily malaise can be traced to the food the individual consumes. I have extensively reviewed available published research material in the hope that I would bring to the reader a comprehensive picture. I did my best in this attempt, and I hope you enjoy reading this book and benefit from the information it contains.

Malaparamba, Kozhikode, Kerala, India   Kodoth Prabhakaran Nair

Acknowledgments

With immense pleasure, I record the many purposeful discussions that I have had with our son, Dr. Kannan Mavila, on this book. Dr. Pankajam Nair, my wife, a nematologist trained in Europe, who left science to become a homemaker, Engineer Sreedevi Mavila, our daughter; and her husband Engineer Arvind, enthusiastically offered many suggestions while writing this book. And I owe a deep debt of gratitude to my many friends, colleagues, past and present; and so many of my former students, scattered all over the world, who taught me many things about life and living a truly purposeful and a happy one. And, finally, to you, the reader, for that leap of faith, in picking up this book, as a validation of the firm belief of a dedicated scientist, trained in soil science, but, having written a book on the science of food, with a totally new perspective as to how it can make or mar the life of a man or a woman. I greatly appreciate the diligent support of Ms. T. Metilda Nancy Marie Rayan during the production of this very important book.

Contents

About the Author

Kodoth Prabhakaran Nair is a globally acclaimed agricultural scientist having worked in different capacities in Europe, Africa, and Asia, for over three decades. A Senior Alexander von Humboldt Fellow, he was the first and only Asian scientist named to the National Chair of the Science Foundation, The Royal Society, Belgium, and was invited by late Nelson Mandela to set up a Faculty of Agriculture in the University of Fort Hare, Alice, Republic of South of Africa, the late President's Alma Mater, from where late Mr Mandela spearheaded the anti-apartheid struggle. He has authored nine books of which one was launched by India's great President late Dr Kalam, in Raj Bhavan, Chennai, India. Currently he lives in Calicut with his nematologist wife Dr (Mrs) Pankajam Nair.

Chapter 1
Introduction

Abstract This chapter traces a historical perspective of food consumption, starting from the Middle Ages, and goes on to discuss an Indian perspective, where food is classified as *satwic, rajasic*, and *tamasic* categories, each imparting a specific mental trait in the individual consuming the food in question and how it reflects on the person's behavioral pattern.

Keywords Middle Ages · Indian Perspective · Health · Behavior

From time immemorial food has been at the center of life, whether for human or for animal. As the common adage goes, some people "live to eat" and others "eat to live." But, the central point is "eating." Food is thought to affect human psyche enormously and, consequently, the behavioral pattern. In politics we have the "dinner diplomacy," the objective being adversaries can always be won over through good food. More pertinently, food plays an important role even in personal relationship. Take for example the man-woman bond, translated into, for example, the husband-wife relationship. It is popularly said that the "way to a man's heart is through his stomach." Reflect for a minute over a tiff between a husband and wife. A smart wife can always smoothen the ruffled feathers of the husband through a good meal.

How does food affect a person's behavior? If one takes the example from the Hindu thought, food can be compartamentalized into three categories, the *satwic* food, the *rajasic* food, and the *tamasic* food. What do these three terms mean? *Satwic (pranic)* food is that which is pure, freshly prepared, not refrigerated, wholly vegetarian, which while being palatable will impart a sense of mental contentment. *Rajasic* food is primarily those which have a meaty origin, spicy, oily, which leads the consumer to quick anger, greed, and invariably to negative mental and animalistic traits. It is not by coincidence that great thinkers like Albert Einstein, consumed wholly vegetarian food. Interestingly, the intelligentsia of India are invariably those who consume *satwic* food. On the other hand, *tamasic* food is stale, sometimes putrid, which makes the person lazy and mentally inert. From the start of recorded history right up to the present, mankind has believed that the food it eats can lead to powerful aftereffects on their behavior. For instance, the ancient Egyptians believed that salt could stimulate passion, onions induce sleep, cabbage could prevent a hangover, and lemon could protect one against the evil eye. The ancient Greeks also thought that diet was an integral part of psychological functioning but added a personality component to the process. They conceived of four temperaments, that is, choleric, melancholic, phlegmatic, and sanguine, which were responsive to heat, cold, moisture, and dryness. Because these characteristics were considered to be inherent properties of food, it was believed that consuming particular dietary items could correct humoral imbalances (Farb and Armelagos 1980).

In the Middle Ages, the belief that there was an intimate connection between the food one takes and his or her sexuality existed, as medieval men and women used food as an attempt to both encourage and restrain their erotic impulses. Truffles, turnips, leeks, figs, mustard, and savory were all endowed with the ability to excite sexual passions, as were rare beef in saffron pastry, roast venison with garlic, suckling pig, boiled crab, and quail with pomegranate sauce (Cosman 1983). To inhibit sexual impulses, lettuce, cooked capers, and diluted hemlock-wine concoctions were sometimes employed, though rarely as often as stimulants (Cosman 1983). Jean Anthelme Brillat-Savarin, the famed French philosopher and gourmand, wrote "Tell me what you eat, and I will tell you what you are" in his treatise *The Physiology of Taste*, first published in 1825. He documented in his treatise several instances of close interrelationship between food and behavior and was the first to document the stimulatory effect of caffeine. He also believed that milk, lettuce, or Rennet apples could gently induce sleep, while a dinner of hare, pigeon, duck, asparagus, celery, truffles, or vanilla could facilitate dreaming. Perhaps this could be one reason why drinking a glass of milk, preferably hot, before going to bed, has become a universal habit with many.

It was in America, in the early years of the last century, that a health reform movement started taking shape, which made people to be aware that food affected mental health, intelligence, sexual prowess, and spirituality. It took more time to spread to the European continent. One of the most prominent leaders of this movement was John Kellogg, known best for introducing the breakfast cereals, in fact the ubiquitous corn flakes, but mercilessly trivialized in the book and movie of the same name, "The Road to Wellville" (Boyle 1984). Kellog widely lectured in the USA, promoting the use of natural foods and decrying the eating of meat, which he believed would lead to the deterioration of mental functioning, while arousing animal passions. Kellog's line of thought was predated by what the Hindu sages thought almost 5000 years ago, who thought that the *Rajasic* and *Tamasic* nature of food one consumes contributed to brutal mentality. There is a reference to meat eating, in the immortal Hindu scripture the *Bhagavad Gita*, what Lord Krishna (the immortal Hindu God) advised Arjuna (the valiant *Pandava* king) that only devils ate dead bodies, of animals. Kellog additionally claimed that the toxins in the body resulting from the digestion of meat produced a variety of symptoms, including mental depression, fatigue, headache, aggression, and mental illness, while spicy or rich food could lead to moral deterioration and acts of violence (Kellog 1888, 1919).

That food and spirituality are linked gave way to the "vegetarian movement" in these regions. Vegetarianism was gaining ground, and millions started giving up meat, as a healthy alternative. The word "vegetarian" is derived from the Latin word

"Vegetus," which means full of vitality, vigor, and cheer. Plant foods signify peace and harmony, are soothing and cleansing, and instill a kind of positivity in one. Meat spells death, blood, harm, pain, and foul smells, all of which create negativity in the one who consumes it. Among these, many turned to vegetarianism out of a firm belief in nonviolence. Knowing pain, and knowing that it is not a good thing, is one of the most elemental forms of nature's intelligence. When juxtaposed to the meat eating habits of the West, it is, perhaps, interesting to speculate that vegetarianism, among the majority of Hindus, in the Indian subcontinent, was a great "civilizational forward leap."

In India, the Brahmins (the uppermost caste in the caste system, descendents of the *Brahma,* the creator) avoid both onion and garlic because they are considered to be *satwic* food. The wide spectrum of objectives which specific foods attain makes thousands of individuals to look at the possibility that a specific type of diet or a specific nutrient can help achieve sexual, emotional, or cognitive equilibrium, apart from the health-giving traits the foods carry. For instance, while fish, especially the oily ones like sardine, carry the omega-3 fatty acid which nourishes the brain, – the British legendary Sherlock Holmes – Dr. Watson duo, where the former's capacity to solve mysteries (murder and what not) was attributed to the generous fish consumption – gooseberry is supposed to bring down blood sugar, and, for those looking for the sexual invigoration, oyster is the prized aphrodisiac. The latter-day product endorsements, especially the advertising gimmick projecting on the TV screen on all kinds of products, for "energy boosting," "checking memory loss," or "cholesterol busting," and what not, leave the consumer totally confused. Product endorsements are quite enthusiastic, and even unreasonably overzealous, where various herbs are now hawked as a natural means for reducing mental depression, while fatty acids in fish oil are meant to enhance human intelligence. Recent years have witnessed a phenomenal increase in the kinds of endorsements that one sees on the internet, television, and magazines, emphasizing the need for "balanced eating," and also, information explosion on the food-behavior relationship. Unlike the past, there is now an abundance of published information on the relationship between nutrition derived from various foods and behavioral patterns, so that one can sift fact from fiction.

The above assumptions need to be taken merely at their face value, since practical experience might lead to contrary results. Alternatively, when more data are gathered, one would have a different emerging view point. To get a holistic view, one also needs to know the methodology scientists follow to arrive at some of the observed changes in human behavioral pattern consequent to the intake of specific food. Though the methodology encompasses primarily nutrition and biochemistry, as supportive fields, one will also need to delve into human psychology, medicine, anthropology, sociology, and public health.

As one would observe in modern medicine, a substantial amount of data is generated through animal studies. And behavioral science is no exception. And when one delves into animal studies, the most important aspect that surfaces is the biosafety. Take, for example, the latest controversy surrounding genetically modified food (GM food). There are lobbies both for and against it. But, on balance, the biosafety aspects are the most crucial when it comes to human nutrition and health. There is a strong perception that GM foods are bio-unsafe. And, much of the available scientific data has been generated through animal studies.

When one decides to pick up this book to read, he or she should also gain an appreciation on the quantum and the act of eating the chosen food that touch one's everyday life, one's behavior, and development. In the developed world, infants are fed immediately by the lactating mother in the postnatal period, while in the developing world, by and large, the infants are dependent on the mother's breast milk for some months, before solid food is given. In fact, in many communities in India, the "rice-giving" ceremony is an important ritual, often performed inside a temple, with the assistance of the temple priest. But, in western civilizations, this is totally unknown. Infants are fed almost immediately after they are born, and children learn of the world through tasting new foods. When extended, families bond through sharing mealtimes, which is a universal practice. More interestingly, in the west, teenagers choose a restaurant for the first date – a practice that is fast catching up in metropolitan cities of India. Adults must work to put food on the plate for the family – a universal phenomenon. In fact, the act of eating provides a script for our lives and, the pleasure and satisfaction one derives from eating might also serve as an end in itself.

Food and behavior is a two-way process, while the former will be influenced by the latter, it will also happen vice versa. What are the factors that lead an individual to the food he or she ultimately chooses to eat is an important aspect of this book. In a global perspective, food preferences are influenced by both cultural and familial aspects against the constraints or otherwise of the economic aspects linked to food preferences. For instance, in rural India, banana would be the preferred choice than apple, though the latter might be available only at a premium, which a poor household cannot afford. Also, the individual's respect for his or her traditions, concerns about health, or fixation on appearance can all interact to determine how that person will interact to determine how he or she behaves around food.

1.1   An Important Indian Perspective on Food and Health

The body is made of five *koshas* (Sanskrit word meaning enclosure) starting from the most basic *Annamaya Kosha* (the word "Anna" in Sanskrit means food; hence, the word means food enclosure). The outermost is *Ananada maya kosha* (the word *Ananda* maya means full of happiness; *Ananda* denotes happiness), and the intermediate enclosure is *Manonmaya kosha* (the word *Mana* means mind, the mind enclosure). This compartamentalization amply proves that if the first food enclosure is filled with pure and *satwic* food, one would end up in happiness, because the route is through the mind. There is increasing medical scientific evidence now to show how chemical toxins build up through food, adversely affecting one's temper and character.

1.2 Food: A Corporate Target

The author, while writing this book, felt that it is also important to note some glaring instances where corporate greed has played an important part in the food habits of people, at large. This has been the scenario in countries such as the USA and Europe, but it is gradually spreading to the developing world, in countries like India. Celebrities, from both the movie world and sports, are lured by huge payments to endorse the food products. The most recent case of the huge, national scandal, in India, with Maggi noodles, where the endorsement was made by one of the most popular cine actor of the country, is a case in point. The product was found to contain beyond permissible levels of the heavy metal lead (Pb), and the company Nestle which was marketing it in India, on a very large scale, had to withdraw all the stocks from the market, hence this short discussion on the topic.

The US sugar industry played a very dubious role in paying the scientists huge sums of money in the 1960s to play down the link between sugar intake and heart ailments and promote saturated fats as the real culprit, instead as shown by the revelations of a historical document. The internal sugar industry documents, recently discovered by a researcher at the University of California (San Francisco campus, UCSF), and published on 12 September 2016 in JAMA (Journal of American Medical Association Internal Medicine), suggest that five decades of research into the role of nutrition and heart ailments, including many of today's dietary recommendations, may have been largely shaped by the sugar industry. "They were able to derail the discussion about sugar for decades," said Stanton Glantz of UCSF. The documents show that a trade group called Sugar Research Foundation, known today as Sugar Association, paid three Harvard scientists the equivalent of US$ 49,000 in today's dollar equivalent (a huge bribe then), to publish a 1967 review of research on sugar, fat, and heart ailments. The studies used in the review were handpicked by the group, and the article, published in the prestigious *New England Journal of Medicine*, minimized the link between sugar and heart health and cast aspersions on the role of saturated fat.

Even though the influence-peddling revealed in the documents dates back more than 50 years, more recent reports show that the food industry has continued to influence nutrition science.

In 2015 Coca-Cola had provided millions of dollars in funding to researchers who sought to play down the link between sugar drinks and obesity, revealed an article in the New York Times. In June 2016 Associated Press reported that candy makers were funding studies which claimed children who eat candy tend to weigh less than those who do not.

The Harvard scientists and the sugar executives with whom they collaborated are no longer alive. One of the scientists was D. Mark Hegsted, who went on to become the head of nutrition in the US Department of Agriculture (USDA). Another was Dr. Fredrick J Stare, the chairman of the nutrition department, Harvard.

In response to the JAMA report, the Sugar Association said that the 1967 review was published at a time when medical journals did not typically require researchers to disclose funding sources. The industry "should have exercised greater transparency in its research activities," the Sugar Association said. Even so, it defended industry-funded research as important. It said, decades of research had concluded that sugar "does not have a unique role in heart disease."

The revelations are important because the debate about the relative harms of sugar and saturated fat continues to this day Glantz said. For many decades, health officials encouraged Americans to reduce their fat intake, which led to many people to consume low-fat, high-sugar foods, that some experts now blame for fueling the obesity crisis in America.

Hegsted used his research to influence the US government's dietary recommendations, which emphasized saturated fat as a driver of heart disease, while largely characterizing sugar as empty calories linked to tooth decay. The fat warnings remain a cornerstone of dietary guidelines in the USA.

References

Boyle, T. C. (1984). *The road to Wellville*. London: Penguin Books.

Cosman, M. P. (1983). A feast of Aesculapius: Historical diets for asthma and sexual pleasure. *Annual Review of Nutrition, 3*, 1–33.

Farb, P., & Armelagos, G. (1980). *Consuming passions: The anthropology of eating*. Boston: Houghton-Mifflin.

Kellog, J. H. (1888). *Plain facts for old and young*. Burlington: Segner.

Kellog, J. H. (1919). *The itinerary of breakfast*. New York: Funk & Wagnalls.

Chapter 2
Conceptual Models in Food-Behavior Relationship

Abstract The chapter discusses, in detail, the conceptual models in food-behavior interrelationships and further examines the role of a nutrition expert.

Keywords Nutrition · Behavior · Ethics

A researcher investigating nutrition may have, as a long-term objective, the promotion of physical well-being. As a short-term objective, one may investigate the role of iron in muscle functioning or building physical stamina. On the other hand, another who wishes to serve the society better would concentrate on mental health. One might come across strange narratives of personal experience, which can motivate the listener to embark on a research effort. For instance, there are some who crave for a chocolate or for a strong drink when the mood is down. The search engines provide the answer. PsychInfo and Medline, which catalogue published research papers which appear in psychological and medical journals, respectively, are extremely useful search engines which give the researcher a lead. For instance, entering the words "chocolate mood," one would find that some studies have indeed been published on chocolate cravings, some on food and mood, and even some on chocolate and mood (Schuman et al. 1987, Macdiarmid and Hetherington 1995). The interesting fact is that one would discover that all these investigations focus on women, who define themselves as "chocolate addicts." One would then wonder if chocolate eating might make a difference in the moods of men. As the reader has already guessed, testing this relatively straightforward proposition involves more than simply handing out Nestle™ chocolate bars to individuals and how they feel. In the following chapter, one would find the methods used by researchers in designing their investigations, and a differentiation between experimental and observational approaches would be attempted. For instance, as an initial test for a hypothesis, conducting a survey might be a useful strategy. Reviewing the articles cited earlier, one would find references to a number of dietary recall instruments, as well as some questionnaires which measure depression. This would help one develop a questionnaire which includes a section on depression and dietary patterns and add some new questions to specifically ask about consumption of chocolate. One could also include some questions which will provide one with demographic information, for instance, age, sex, nationality etc., though one must be careful not to ask any questions that would permit identification who a specific respondent is. This would be an infringement on the individual's privacy. If a researcher doing this survey is affiliated to a university, in particular, in the USA or in Europe, the investigation would require approval by the university's Institutional Review Board. Such stringent institutional stipulations, as one commonly observes in these regions, might not be stringently followed in many institutions in the developing world, though, at the departmental or individual level, these stipulations would be enforced. In the case of an anonymous survey, where one cannot identify the respondent, and the participation of the respondent is voluntary, an exemption would be granted from the abovementioned stipulation by the concerned host institution.

When an investigation is conducted that relates to human subjects, ethics forms the focal aspect of the investigation. The cardinal principle that applies to any investigation, where humans are involved, is one concerning ethics. There were instances of flagrant violations of this principle during the World War II effort of Nazi Germany. Instances such as feeding extracts of dried plant juice through its addition to the flour to the general population to make women sterile and exposing men to X-rays without their knowledge with the same objective are such examples. More shockingly, war prisoners in the Buchenwald concentration camp in Germany were injected with typhus fever virus in an effort to develop a vaccine against typhoid. After the war trail and through the efforts of the Nazi War Crimes Tribunal, fundamental ethical principles for the conduct of research involving humans were generated and made part of the Nuremberg Code, issued in 1947, which sets forth a number of conditions that must be met before research involving human subjects is ethically permissible. Yet, one observes the blatant violations of some of these principles in the pharmaceutical sector, in a country like India, where unscrupulous doctors conduct clinical trials on new products in rural parts of the country, without the knowledge of the patients, generally very poor. Pecuniary gain is at the center of these flagrant violations. Primary among the tenets of the abovementioned code is that voluntary consent of the human subject is absolutely essential, with additional elements concerning safety considerations in designing, conducting, and terminating the experiment.

2.1   Some Flagrant Violations of the Ethics Code

The infamous Tuskegee Syphilis Study, where black men with syphilis were left untreated to track its effects, is a flagrant, yet sobering violation of the Nuremberg Code. The study was initiated in the 1930s without the men giving their informed consent. Shockingly, when the antibiotic penicillin was made available in the 1940s, the men were made neither aware of this nor treated with the antibiotic. In fact, the study continued until 1972, when accounts of it finally appeared in the American press. It was discovered that a similar study was conducted in Guatemala from 1946 to 1948 by the US Public Health Service, without obtaining an informed consent from the patients involved. This shocking and shameful discovery was universally condemned, and President Obama had in 2010 publicly apologized to the Guatemalan people for what he called "a crime against humanity." There are other examples of atrocious violations of ethics in research. From 1946 to 1956 mentally retarded boys at the Fernald Sate School in Massachusetts, USA, who thought they were joining a science club, were fed radioactive milk with their breakfast cereal to learn how radioactive iron and calcium were absorbed by the digestive system. In 1963 investigations were undertaken at New York City Chronic Disease Hospital to gather information on human transplant rejection process. Patients in varying degrees of debilitation were injected with liver cancer cells, hoping that their bodies would reject these cells. Examples can be multiplied. These shocking revelations in the USA led to the formulation of the National Research Act (PL 93–348) in 1974 formulating strict stipulations to be enforced while undertaking research with human subjects. The National Commission for the Protection of Human Subjects of Biomedical and Behavioral Research to identify the ethical principles which should guide the conduct of all research involving humans was thus established. The Commission's efforts led to the formulation of a document known as The Belmont Report: Ethical Principles and Guidelines For the Protection of Human Subjects, published in 1979, has the following provisions:

> Requiring every researcher to respect the autonomy of the individual connected with the investigation, where the information gathered is through informed consent
> Treating the subjects investigated in an ethical manner
> Ensuring that the selection of the human subject is done with justice to ensure that a benefit to which a person is entitled is not denied without good reason or a burden is imposed unduly.

Of late, there have been many instances, in a country like India, where nutrition and/or food item endorsement by a "celebrity," mostly by cine actors, for a huge price, that is, is cloaked in misguiding information. The classic case of the Maggi noodle, (a Nestle product widely marketed in India), which was endorsed by a leading cine actor, where the product was suspected to contain levels of monosodium glutamate (MSG) much above permissible limits, endangering human health, amounts to, in fact, nutrition quackery and psychological misconduct. Clinical psychologists, whose very profession is built upon the application of their training to assist individuals in achieving mental health, are obviously bound to a code of conduct which mandates trust between the therapist and the client. Different psychological therapies exist, but, to become a psychologist, one has to master graduate work in psychology, complete an extensive internship wherein his or her training as a therapist is supervised, and finally pass an examination for license. The rigor of such training can vary from country to country. One psychologist's mode of therapy with a particular client may be less effective than that of a colleague, but no blame is attributable given the psychologist's disclosure of alternative approaches and the client's free choice. But, the psychologist must terminate a consulting or clinical relationship when it is reasonably clear that the client is not benefiting from it. In the USA, adherence to Ethical Principles of Psychologists is demanded, and a psychologist can have his or her license suspended if he or she is found to have violated any of these principles. For example, disclosing personal information about a client to someone other than another psychologist concerned with the case or for whom the client has given permission, would be ground for suspension. For that matter, any information obtained about a research participant during the course of an investigation is also considered confidential, with the researcher responsible for assuring anonymity in any report of the investigation. Such professional rigor is not adhered to in many instances in a country like India.

2.2 Where Does a "Nutrition Expert" Fit in the Whole Scenario?

Coming back to the main topic of this book – "Food and Behavior" – it is important to examine where does nutrition, the backbone of food, fit in. In fact, a nutrition credential does exist in the form of a Registered Dietitian (RD). The RD designation indicates that the individual has completed at least a Bachelor's degree from an accredited college or university, has mastered a body of nutrition-related courses specified by the American Academy of Nutrition and Dietetics, has successfully completed a supervised work experience, and has passed a national qualifying examination. The Academy has its own Code of Ethics for an RD. Hence, an RD is clearly the nutrition professional concerning the health aspects of nutritional sciences. In case one is seeking sound nutritional advice concerning one's own health, it would be advisable to seek the help of an RD. Next to this preference, it would be one's general physician, who would have studied human physiology, biochemistry, etc. during the medical course, who would be the next one to choose from. These stipulations apply to the situation in the USA only. But, the reality is that anyone can call himself or herself a nutrition expert. In fact, there are far more people calling themselves nutritionists than there are RDs.

The field of nutrition is broad enough that quackery presents itself as a real danger. The word "quack" is derived from *quacksalver*, which is an archaic term used to identify a salesman who quacked loudly about a medical cure, such as a lotion, salve, or, in today's world, the wonders of a particular nutritional product or regimen. Nutrition quacks may attempt to market their alternative products or regimen to anyone, but there are certain populations which would be the potential target. And this consists of those patients who suffer an incurable disease like cancer or AIDS wherein the victim is desperate to have a cure at any cost and may willingly fall a prey to these unscrupulous people who peddle their ware.

References

Macdiarmid, J. I., & Hetherington, M. M. (1995). Mood modulation by food: An exploration of affect and cravings in "chocolate addicts". *British Journal of Clinical Psychology, 34*, 129–138.

Schuman, M., Gitlin, M. J., & Fairbanks, L. (1987). Sweets, chocolate and atypical depressive traits. *Journal of Nervous and Mental Disorders, 175*, 491–495.

Chapter 3
Nutrition-Behavior Interface: The Way Forward in Research

Abstract The chapter discusses, at great length, experimental designs in devising nutrition-behavior interface.

Keywords Experimental designs · Independent and dependent variables · Comparative merit of experimental designs

The nutrition-behavior interface arises from an array of research designs. For instance, while an epidemiological approach could be selected by a team of nutrition specialists, a group of psychologists are likely to choose a cross-sectional design to investigate changes with development. The former is exemplified by an investigation where the daily consumption of fruits and vegetables in a school with children of elementary school-going age is related to the most recent score obtained in the spelling test. For the latter, the researchers compared the same effect for the classes of the first, third, and fifth graders in the same school. Both of these approaches would tell the investigators something about the relationship between dietary patterns and academic achievement, but one approach would be chosen over the other depending on what specific question the researchers were most interested in answering. Notwithstanding the marginally different tactics illustrated in the above examples, in actual practice both the research groups share a good number of research approaches, which is not surprising, because both groups follow the scientific method. Despite the wide variety of research designs, most can be subsumed under one of the two general strategies, the experimental versus the correlational approach as detailed in the following (Table 3.1).

In correlational designs, there are, normally, two variables, one an independent variable and another a dependent variable. Correlational designs are very appropriate to investigate physical relationships, like the temperature-pressure interrelationship, where a specific increase in the pressure results in a *direct* and *proportional* increase in temperature of physical entities. In biological systems, however, such studies will only show a general directional relationship, but cannot be taken as absolute **per se**. Such correlational investigations between two related variables are known as "simple correlations." The "b" value in the regression function $Y = a + bx$

Table 3.1 Comparative merit of experimental designs

Correlational designs	Experimental designs
Case control	Dose response
Cross-sectional	Quasi experiment
Exploratory	Dietary challenge
Epidemiological	Crossover design
Longitudinal	Non-equivalent groups
Retrospective	Dietary replacement

Note: A fool-proof experiment is the **sine qua non** of scientific research. The details in the above table are merely "directional"

refers to the slope of the equation, where the variation in the "independent" variable "x" has a direct and proportional increase in the dependent variable "Y," if it is a study between temperature and pressure, as explained above. The "r" value in the correlation shows the degree of "co-relationship" between "x" and "Y." Squaring "r" gives the "coefficient of determination" expressed in percentage. In other words, it would show the extent of influence of x on Y. However, there are multiple correlations or curvilinear relationships, where more than two variables are under scrutiny. Generally, these relationships are expressed as a "correlation coefficient" ("r") value. In a correlational investigation of the relationship between variables, the researcher would operationally define the variables of interest and then observe or obtain measures of the relevant variables. Supposing an investigator starts an investigation on the interrelationship between sugar and hyperactivity, he or she might ask a group of children to complete a diet record of everything they ate over a weekend and also ask their parents to report on how active their children were at bedtime, using a standardized activity rating scale. The researcher would then determine the total sugar content of the children's diets using a packaged computer program. Currently available programs can accurately and rapidly calculate energy intake, as well as grams ingested, for sugar and assorted micro- and macronutrients. Similarly the children's scores on the activity scale could be easily tabulated with higher scores indicating higher activity levels. Finally, the researcher would run a correlational analysis to learn if there is a relationship between sugar intake and activity level – is it really the case that the children who ingested more sugar also displayed higher activity level?

From the abovementioned investigation, it would be tempting to conclude that there exists a positive relationship between sugar intake and activity level, if a positive correlation coefficient was found. However, such a conclusion would be premature for several reasons, some specific to this particular investigation but others problematic to all correlational studies. In an investigation like the abovementioned one, there are several conditions that must be met, before one can accept the validity of the results.

To ensure that the investigation is scientifically fool proof, certain parameters must be strictly ensured. First of all, valid measures of nutrient intake must be obtained. These would constitute diet history, food record, food frequency questionnaire, and 24 h diet recall. The 24 h diet recall is one of the standard approaches to measure dietary intake in which subjects are asked to record every-thing they have eaten over the previous day. There are wide day-to-day variations in any individual's food intake, so one 24 h record is unlikely to provide an accurate determination of average dietary patterns. Some experts suggest a 7-day minimum record to be kept while running correlational studies (Anderson and Hrboticky 1986). The accuracy of children's dietary records of their meals, snacks, and portion sizes leaves much to be desired, which will, inevitably, lead to erroneous conclusions.

Secondly, appropriate sampling techniques must be used to minimize extraneous variables that might affect the behavioral outcome. By sampling, one is referring both to the subjects that comprise the group of individuals who are investigated and to the segments of behavior that are used to represent the variables of interest. In general, a larger number of subjects (referred to as "n" – sample size) are preferred relative to a smaller number. If the n size is too small, the probability of observing relationships between a nutrient variable and a specific behavior is reduced, and it may be falsely concluded that no relationship exists. This is because the statistical procedure applied to analyze the data will detect significant variations between treatments, only if the sample size is large enough. When the sample size is small, even if there are significant differences between treatments, they would not be shown as statistically "significant." On the other hand, correlational studies which involve a huge n size risk the possibility of finding "false positives," which means associations that appear to be present may just be an artifact because of the large n size. For example, when correlations are run between several nutrition variables and a behavioral measure, the chances of achieving statistically significant results increase. In epidemiological investigations even small correlation coefficients can attain statistical significance with large n size, which makes the researcher to be careful in determining whether or not the association is conceptually meaningful (Anderson and Hrboticky 1986).

Apart from the population size (in the above-discussed example of the children), the sampling of behavior is also a concern. In the study, the sugar intake measure-ment was done on a 2-day-long dietary period, which may or may not be adequate. But, the parental ratings of activity might have been colored in some cases by the perception of the parents of their child's general activity level. This means they may be overestimating the child's activity at bedtime based on their general perceptions from years of living with the child and not on their bedroom display. Ratings of activity may be particularly subjective, but ratings of sociability, aggressiveness, shyness, helpfulness, and, in fact, anything which may be personality based will all depend on how much time the subject has been observed. Human behavior can vary from day to day, hour to hour; hence, while bedtime might have been a sufficiently reasonable index of the child's activity level, it may not be enough to condemn or acquit sugar if its effects are short-lived. Whenever behavior is being observed and rated, one has to address whether enough behavior patterns have been recorded to make a valid estimate of what is typical of that person's behavior.

Thirdly, correlational studies cannot establish, with absolute certainty, causality. In the case discussed, above, one cannot conclude, for sure, whether bedtime hyperactivity resulted from enhanced sugar intake or higher sugar intake led to bed-time hyperactivity. In other words, children who are more active might crave more sweets and candies than others who are sober. To establish causality, one should be certain which preceded which. Additionally, an important third variable is exposure to visuals like television. The children might have seen several television visuals where sweets' consumption would have been shown through endless attractive commercials for candy, cookies, other sugary snacks, etc. Also, visuals like jumping, dancing, and kickboxing could have led the children to imitate what was seen on the television screen.

The problem is not solved even by correlating television watching and sugar intake. Again, measuring television watching at approximately the same time as diet and activity were observed would imply that these variables varied simultaneously. Another contributory factor could be undisciplined parenting. Notwithstanding all these limitations, a correlational study would still provide a lead in association between variables and could lead to the formulation of a hypothesis about the nature of the relationship and subsequent testing using an experimental design.

The correlational studies apply to epidemiological aspects as well. For example, a psychologist who counsels young women with anorexia nervosa may be trying to identify factors in her family history that are relevant to her current condition. He would do this by starting with an individual who was diagnosed with an eating disorder and then examine elements of her past that seemed to characterize her family of origin. If enough of his clients disclosed similar memories of growing up, the psychologist might infer a relationship between parental hostility, for example, and the daughter's disorder. Despite the appeal of such a correlation, he would have to recognize that other factors might be the actual cause of his client's eating difficulties.

Two additional designs, popular in developmental psychology, can answer queries about changes in behavior accompanying age. The cross-sectional approach surveys the rich, middle class, and poor among the population to draw conclusions about people's lifestyle as a function of socioeconomic status. There are numerous studies of this nature, where the eating habits of social groups by income, ethnicity, or geographic region are of interest. In behavioral sciences, however, the cross-sectional design refers to a study in which subjects of different ages are observed in order to determine how behavior may change as a function of age.

3.1 Experimental Approaches

In contrast to the correlational investigations, experimental designs have the potential to identify causal links between diet and behavior. Where a correlational study may include a number of variables, equivalent in value until associations are determined, an experimental study includes two specific types of variables. The independent variable manipulated by the investigator constitutes the treatment that the subjects receive. It may be the addition of a specific nutrient to the subjects' diets or the application of a nutrition lesson to a class of students, but in either case the investigator is controlling what the subjects are exposed to or how their experience within the investigation differs. The dependent variable refers to the observation, measure, or outcome that results from the manipulation of the independent variable. If the manipulation of a specific dietary component (the independent variable) significantly alters the form or magnitude of a behavioral measure (the dependent variable), a causal relationship can be then postulated.

Though various investigations on nutrition-behavior relationship can vary widely in terms of their complexity, certain common elements will characterize any sound experiment. To begin with, an experiment will include a minimum of two groups of subjects. The group of greatest interest, the focus of the investigation, is referred to as the experimental or treatment group. This group of subjects receives the treatment which is the independent variable. The second is referred to as the control group. In the simplest case, the control group receives nothing and serves a standard for comparison purposes. Any difference that is seen between the treatment group and control group at the end of the experiment can therefore be attributed to the independent variable. To be certain that such differences are due to the independent variable alone, it is important to ensure that the groups are similar in all respects prior to the application of the treatment. This is, more often than not, difficult to ensure.

In studies where the participating subjects have reason to expect that their receiving a treatment might change their behavior or feelings of well-being, a placebo control group may also be included. The placebo effect refers to the phenomenon that when human beings take any treatment that they believe will be beneficial, their chances of improving are increased (Hrobjartsson and Gotzche 2001). Because food is a part of so many aspects of everyday life, it holds significance far beyond its nutritional value. Research on nutrition and behavior is particularly susceptible to a subject's belief that specific nutritional supplements may improve his/her disposition and performance. To contain this false effect, an experimenter interested in the effectiveness of St. John's wort in relieving depression could give to the experimental group the actual treatment (pills containing the herb), to a placebo group a mock treatment (sugar pills), and to the control group nothing. If the herbal treatment is effective, following the experiment, depression scores should be the lowest for the treatment group. Investigators often harbor biases about the expected outcome of a treatment, since, after all, they too are human, given to human fallacy, and so the experiment should be conducted under double-blind conditions. Under such a condition, neither the experimenter nor the subjects know whether the subjects are receiving the treatment or placebo. If this is not ensured, the entire exercise would turn out to be a meaningless one.

3.2 How to Design an Experiment?

How to design an experiment? The next question is: How would one design an experiment where the causal effect is properly examined? Recall the earlier details of the experiment that was conducted on sugar intake and hyperactivity among children. First, one must decide on an age range on which to focus the attention. With the permission of the Board of Education, the children's parents, and the children, the following procedure can be implemented. One must understand that such procedures are applicable only to developed countries like the USA. On a given Wednesday (pizza day in the cafeteria, as in the USA), the experimenter arrives just before the lunch and announces to the class that as a special treat that day, in addition to their fat-free milk, each child would be given a 12-ounce bottle of root beer. After the children are seated in the cafeteria, each child is handed a bottle. It will have the manufacturer's name covered with special labels of yellow or green which include the school's name and picture of their mascot. Each child receives a yellow or green bottle. Though the teacher or the students are not informed beforehand, yellow bottles contain diet root beer (no added sugar) and green bottles contain regular root beer with added sugar (40 g). Each child's name is noted and also the color of the bottle received by the child. The experimenter observes that each child ate a slice of pizza and drank all the root beer, though some did not finish their salad or milk. Following the lunch, the child goes for 20 min recess to the playground and we ask the teacher to observe each child. After the recess, the teacher completes the activity rating scales on each child while the children are read a story by the researcher.

If the experimenter finds that the activity scores of the children who had the regular root beer (treatment group) is slightly higher than those in the case of the children fed the diet root beer (control), one is led to conclude that sugar intake invariably enhances activity. Despite the expected finding, why must one be careful in avoiding a generalization that sugar intake will, without doubt, lead to higher activity? It is for the following reasons:

The assumption that all the children had the *same* previous history (in the day) before the treatments were imposed might simply be false. For instance, some children might have been better nourished than the others before they drank the root beer. Yet, some others might have skipped breakfast, some others might have slept more, and some might have come from single parent families. However, it can be expected that these differences have been randomly dispersed across the groups. While a correlational study benefits from a larger sample size, equivalent subsamples are a greater concern in an experiment. The target class might have been too small to ensure that individual differences in children were evenly distributed across the two halves.

Possibly, 40 g sugar was inadequate to reflect a discernible activity change. Maybe 60 g is the threshold level of sugar necessary to demonstrate a behavioral effect.

Perhaps a steady diet of highly sugared food does increase activity, unlike a single drink. As behavioral effects may only appear with prolonged exposure, both acute and long-term studies should be used to assess nutrition-behavior relationships.

Time of the day that a nutrient is tested may possibly influence its behavioral effects. Diurnal rhythms of alertness may be different between morning and afternoon, and a greater effect may have been seen if the sodas were distributed at snack time than during lunch time. Time is also relevant to the behavioral outcome, as immediately after lunch may have been too short an interval in which to see an effect on activity level.

The greater nutritional context of any dietary manipulation must be considered when designing an experiment. What the children had for breakfast may have varied from child to child, which is why some studies require that the subjects fast before eating a test meal. Random assignment should have controlled this. But, altering one dietary variable can often alter the intake of others, as with low-fat foods that contain more sodium. Although the use of root beer as opposed to a cola beverage eliminated caffeine (root beer has no caffeine; regular cola has), serving the soda with lunch as opposed to on an empty stomach may also have made a difference; rather, it reduced the differences between groups that could have been found.

Notwithstanding the above weaknesses, the experimental approach is still the best bet to demonstrate causality. As with correlational studies, a number of variations also exist. The above example is commonly referred to as dietary challenge study. In such an experiment, behavior is usually evaluated for several hours after the subjects have consumed either the substance being studied or a placebo. This approach is also referred to as a "between subjects design."

In certain nutrition-behavior investigations, dose-response procedures are sometimes used. For instance, instead of a low dose (in the example cited above 40 g sugar), multiple doses of caffeine (0, 100, 200, 300, and 400 mg) may be tested to determine if there is a systematic relationship between that dietary variable and attention (Brunye et al. 2010). The lack of a systematic relationship may be a warning that the apparent effect is spurious or that the variability is greater than expected.

The behavioral effects of two diets – where one contains a food component of interest and the other is similar as possible to the experimental diet except for that component – are compared that constitute a diet replacement investigation over a period of time. For instance, the fat content of a regular diet could be replaced with a fat substitute in order to determine if subjects will compensate for the reduction in calories by eating more food (Bray et al. 2002). Such a manipulation would be relatively innocuous if done over a day or two, but differences in energy intake over a number of months could be attributed to a change in perceived hunger due to the experiment. An obvious advantage of dietary replacement studies is that chronic dietary effects can be examined. However, it is often difficult to make two diets equivalent, except for the food component that is being investigated, making double-blind techniques relatively hard to employ.

When a characteristic or trait which cannot be manipulated is the variable of interest, a quasi or naturalistic experiment is conducted. As an example, one would not deliberately deprive children of iron to observe the behavioral effects of anemia. However, one could identify children who were alike on a number of variables like age, except for the presence or absence of iron deficiency, and compare their performance on a battery of cognitive tests (Lozoff et al. 2006). In such a study, iron deficiency would be viewed as the independent variable.

3.3 Nutrition-Behavioral Research: The Role of Independent and Dependent Variables

The important question that concerns the nutrition-behavioral research inasmuch as the independent and dependent variables are concerned is: Is there a logic to which variables may be looked at? Earlier in this book, a statistical view of independent and dependent variable, their co-relationship, level of significance, etc. were explained. That is, are there certain "usual suspects" that most scientists would find intriguing? As shown below, there are a good number of variables that investigator would find appealing. The following (Table 3.2) details variables that are typically tested in studies of peoples' eating behavior, when using an experimental design (Rodin 1990).

Among the independent variables, weight, gender, personality, and diet history may be considered as organismic or individual factors, but greater attention is bestowed onto setting variables and characteristics of the food itself. Social and cultural factors refer to regionalism in dietary habits, such as deep-fried foods in the US southwest or pasta families of the Italian heritage. When it comes to countries in South Asia, like India, the focus would be a lot more on rice-based foods. Down south in China, it would be starch-based foods like noodles, etc. The setting or context factor may be illustrated by the phenomenon of people eating and drinking more when they are in a social situation like a party. External cues like the clock than any real feeling of hunger characterize eating at noon on a Saturday or after having had a late breakfast because of oversleeping. Buffet-style restaurants which are increasingly becoming popular in the developing countries like India offer wide choices. Cognitions refer to our ideas about food, such as what is food in the first place. In certain cultures, insects, eyeballs, or even dog meat might be considered palatable or even a delicacy, while in most western cultures and even in countries such as India, such food would invite disgust. Palatability, of course, refers to ease of acceptance. In Europe, for most children, a mild cheddar is likelier to be preferred over a mottled blue cheese. Food cravings that are based on a deficiency or a perceived need for some quick calories may be considered nutrient-related or energy density variables, respectively. Finally, some diet and behavior researchers have investigated how other food characteristics like volume and texture and whether it is presented in a solid or liquid form may influence consumptive behavior.

Table 3.2 also carries details of the dependent variables that diet and behavior researchers find most interesting. Some may be measured through observational

Table 3.2 Variables in diet and behavior research

Independent variables	Dependent variables
Organismic or individual characteristics	Amount of food consumed
Social and cultural factors	Eating rate
Setting and context	Manner of eating
External cues	Eating frequency
Cognitions about food	Motivation for food
Palatability of food	Physiological responses
Energy density of food	Judgment of food quality
Texture and volume	Hedonic rating of food
Liquid or solid form	Satiety feelings

After Rodin (1990)

methods. For instance, the amount consumed or the rate of eating that the subject exhibits can be video recorded for later coding. The manner in which the subject approaches the meal, eating the greens first or slicing the meat throughout even before taking one piece, refers to the microstructure of food intake. When one extrapolates the situation to the orient, there would be a world of difference. For instance, the Chinese and Japanese would be using the chopstick, and this would change the entire scenario. In India, where many prefer to eat with the hand, using fingers then again the scenario would change vastly. Also in the Middle East, eating with hand is preferred to the use of knife, fork, and spoon.

How the individual spaces the meals or snacks denotes his or her frequency of ingestion. For example, after listening to a Public Service Announcement on Health in the USA, a subject might prefer to eat a salad rather than choose the usual steak, which is a departure from the usual motivation for food. More sophisticated observation can be made of a subject's physiological responses, such as an increase in the heartbeat rate after ingesting caffeine or perspiring after some hot Szechuan food. The latter experience might simply be not there with an Asian subject used to eating spicy food. While such objective measures may be preferred, asking the subjects themselves about their perceptions can also advance our knowledge of food and behavior connections. For example, how much did taste or texture affect their enjoyment speaks about their own judgment of food quality. How good they perceived the food to be can be assessed via hedonic ratings. And finally, do their ratings of hunger or satiety suggest that the test meal was adequate to promote contentment? The model shown below (Fig. 3.1) illustrates the multitude of factors that are considered when an individual makes a choice of a specific food (Furst et al. 1996).

The term "Who" constitutes characteristics which refer to the individual, either descriptive like age, sex, etc. or biological like heredity, well-being, etc., or personality-based traits like depression, activity level, etc. The term "Where" relates to the physical environment, like time and place of food choice, and cultural norms that influence the individual's decision-making capabilities. The "Why" refers to food perceptions which relate to the individual's food choices as belief or sensory-based, as opposed to hunger cues (Krondl 1990). From the above pictorial represen-

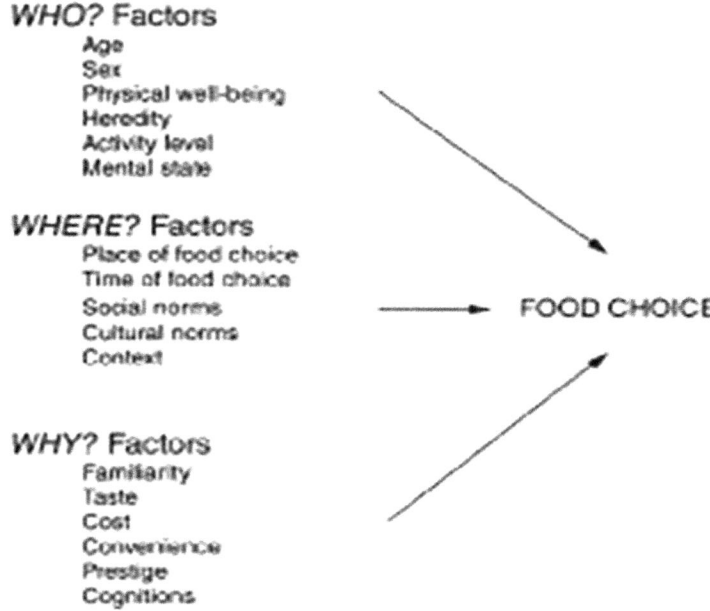

WHO? Factors
Age
Sex
Physical well-being
Heredity
Activity level
Mental state

WHERE? Factors
Place of food choice
Time of food choice
Social norms
Cultural norms
Context

FOOD CHOICE

WHY? Factors
Familiarity
Taste
Cost
Convenience
Prestige
Cognitions

Fig. 3.1 Food selection behavior model. (Source: Krondl 1990)

tation, one can conclude that there is a lot that goes into what makes a good meal at home or making selections in a restaurant, than merely knowing the characteristics of the food that is available.

The limitation of the preceding list of independent and dependent variables, given the perspective on nutrition and behavior, is that while they all have something to do with eating, they seem to go no further. In all fairness it must be mentioned that both Rodin (1990) and Krondhl (1990) models were developed to assist in our understanding of a large body of work in diet and behavior that focuses on the choice of food and dietary intake. Indeed, eating behavior is the stuff of which many diet and behavior investigations are based on, and what a person eats has at the very least indirect effects on other behavioral patterns. Dietary components may influence mood states, activity levels, social interactions, cognitive performance, as well as a host of other behavioral variables. While the "Who, Where, and Why" formulation may be useful in predicting the diet on which an individual may subsist, in the following chapters, far more behavioral outcome patterns will be discussed, than simply what foods are chosen or how much is eaten.

References

Anderson, G. H., & Hrboticky, N. (1986). Approaches to assessing the dietary component of diet-behavior connection. Nutrition reviews. *Diet and Behavior: A Multidisciplinary Evaluation, 44*, 42–51.

Bray, G. A., Lovejoy, J. C., Most-Windhauser, M., Smith, S. R., Voaufiva, J., Denkins, Y., deJonge, L., Rood, J., Lefvre, M., Eldridge, A. L., & Peters, J. C. (2002). A 9-mo randomized clinical trial comparing fat-substituted and fat-reduced diets in healthy obese men: The Ole Study. *American Journal of Clinical Nutrition., 76*, 928–934.

Brunye, T. T., Mahoney, C. R., Lieberman, H. R., & Taylor, H. A. (2010). Caffeine modulates attention network function. *Brain and Cognition, 72*, 181–188.

Furst, T., Connors, M., Bisogni, C. A., Sobal, J., & Falk, L. W. (1996). Food choice: A conceptual model of the process. *Appetite, 26*, 247–265.

Hrobjartsson, A., & Gotzche, P. C. (2001). Is the placebo powerless? An analysis of clinical trials comparing placebo with no treatment. *The New England Journal of Medicine, 344*(21), 1594–1602.

Krondl, M. (1990). Conceptual models. In G. H. Anderson (Ed.), *Diet and behavior: Multidisciplinary approaches* (pp. 5–15). London: Springer.

Lozoff, B., Jimenez, E., & Smith, J. B. (2006). Double burden of iron deficiency in infancy and low socioeconomic status: A longitudinal analysis of cognitive test scores to age 19 years. *Archives of Pediatrics & Adolescent Medicine, 160*, 1108–1113.

Rodin, J. (1990). Behavior: Its definition measurement in relation to dietary intake. In G. H. Anderson (Ed.), *Diet and behavior: Multidisciplinary approaches* (pp. 57–72). London: Springer.

Chapter 4
The Brain-Behavior Link: A Conundrum

Abstract The chapter discusses, at length the role of the Central Nervous System (CNS) and its cruciality in behavioral pattern, the development of the CNS, development of the brain, effects of lipids and fatty acids, effect of polyunsaturated fatty acids and early behavior, inter-relationship between infant development and essential fatty acids (EFAs), the eternal question of breast milk vs. formulas comparison, and adult cognitive capability linked to the long-chain polyunsaturated fatty acids (LCPUFAs), Cholesterol linked adult behavior, cholesterol-antisocial behavior link, and cholesterol-serotonin link.

Keywords Brain · Central nervous system (CNS) · Long-chain polyunsaturated fatty acids · Early behavior · Breast milk · Formulas · Essential fatty acids · Adult cognitive capability

The brain-behavior link is, truly speaking, a conundrum. Changes in the functioning of the central nervous system (CNS) will ultimately lead to alterations in behavior, in both humans and animals. In other words, whatever affects the brain affects the behavior. Diet influences the chemistry and function of the brain. Although large gaps in knowledge exit at the biochemical, physiological, and behavioral levels in terms of our knowledge of the precise effects of nutrition, on the function of the brain, we do know that diet exerts an effect on the development and maturity of the brain. Indeed, many constituents of the diet, such as minerals, vitamins, and macro- and micronutrients, have long been shown to influence the brain function. In recent decades, research has further determined that essential fatty acids, as well as certain amino acids, also play a role in brain development and function. In the case of fatty acids, functional effects are evident, but the underlying mechanisms are poorly understood, while the process of influence for amino acids is well documented, but their consequences are not yet fully fathomed (Dangour and Allen 2013).

4.1   The Central Nervous System (CNS) and Its Cruciality in Behavioral Pattern

Since the CNS is central to our understanding of brain and behavior connection, one should start with a description of its structure and development. The brain is one of the two major components that comprise the CNS, the other being the spinal cord. The adult human brain weighs from 1.3 kg to 1.4 kg (about 3 lb), while the spinal cord is approximately 44 cm in length (about 17 in). Neurons, or nerve cells, comprise about half of the volume of the brain and form the structural foundation for the brain. Neurons serve as the information processing and transmitting elements of the CNS. Their capacity to process and transmit information depends on their ability to generate and conduct electrical signals as well as to manufacture and secrete chemical messengers. These chemical messengers are known as neurotransmitters. Recently it has been estimated that the adult human brain contains about 86 billion neurons (Azevedo et al. 2009) though some have put it as high as 100 billion (Chanhangeux and Ricoeur 2000). Neurons are similar to other body cells. They have a nucleus, containing genes, contain cytoplasm, mitochondria, endoplasmic reticulum, and other organelles, and a membrane surrounds them. However, they are different from other cells in that they possess special properties which allow them to function as the components of a rapid communication network.

Their ability to communicate is facilitated by them having specialized projections called dendrites and axons, them forming special connections called synapses, and them producing special chemicals called neurotransmitters. Like snowflakes, no two neurons are identical but most share certain structural features that make it possible to recognize their component parts – the soma, the dendrites, and the axon (Figs. 4.1 and 4.2).

The soma or the cell body contains the nucleus of the neuron, much of the biochemical material for synthesizing enzymes, and other molecules necessary to

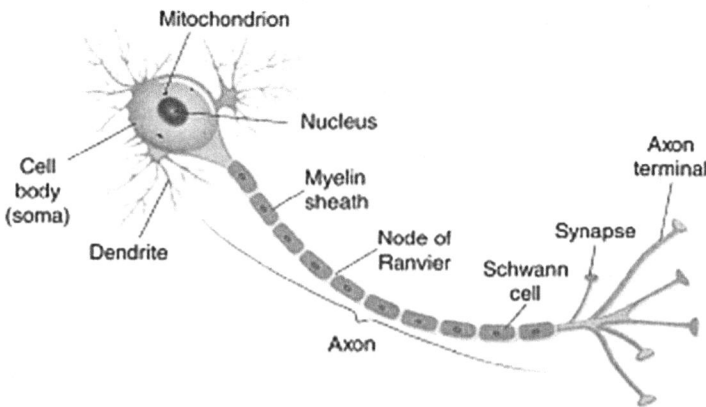

Fig. 4.1 Structural configuration of a neuron. (*Source:* Shutterstock)

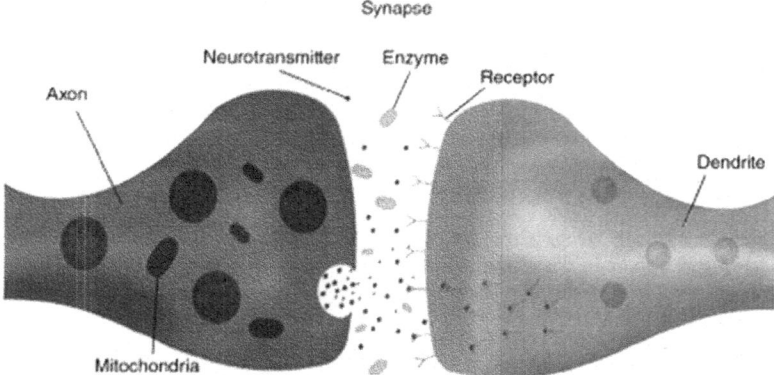

Fig. 4.2 Structural configuration of a synapse. (*Source:* Shutterstock)

ensure the life of the cell. The dendrites are fine extensions which branch out to form a treelike structure around the soma (the word dendrite derives from the Greek word for tree, i.e., "Dendron"). Dendrites serve as the main physical surface through which the neuron receives incoming information from other neurons, meaning that the dendrites bring information to the cell body. The surface of the dendrite is somewhat rough or corrugated, with outgrowths known as spines which receive the chemical messages.

Each dendrite receives messages from hundreds of other nerve cells which will affect the activity of the neuron. As a result of these competing messages, the information that the neuron receives may or may not be transmitted down its axon to another nerve cell. The axon extends from the cell body and provides the pathway over which signals travel from the soma to other neurons. In other words, the axon takes information away from the cell body. Axons are usually thinner and longer than dendrites and exhibit different types of branching pattern. While the branches of the dendrites tend to surround the cell body, the branches of the axon occur near the far end away from the soma where the axon will communicate with other neurons.

The point at which information flow takes place from one neuron to another is referred to as a synapse (derived from the Greek word "syn-haptein" or "together to clasp"). The synapse is a small gap which separates two neurons and consists of a presynaptic ending (that contains neurotransmitters, mitochondria, and other cell organelles), a postsynaptic ending (that contains receptor sites for neurotransmitters), and the synaptic cleft (a space between the presynaptic and postsynaptic endings, Fig. 4.2). Each of the branched ends of the axon is enlarged and forms a terminal bulb known as a bouton. The bouton contains numerous small spherical structures called synaptic vesicles which hold the chemical messenger, that is, the neurotransmitter.

For neurotransmission between nerve cells to occur, an electrical impulse first travels down the axon to the synaptic terminal. At the presynaptic ending, the electrical impulse triggers the migration of vesicles (containing the neurotransmitters) toward the presynaptic membrane. The vesicle membrane then fuses with the presynaptic membrane and its neurotransmitters are released into the synaptic cleft. The neurotransmitter molecules then diffuse across the synaptic cleft where they will bind with the receptor sites of the postsynaptic ending. When a neurotransmitter binds to a receptor at the dendritic spine of another nerve cell, the electrical response of the receiving neuron is affected. The neurotransmitter can either excite or inhibit the postsynaptic cell, which will determine its action potential. If the excitatory postsynaptic events are numerous enough, they will combine to cause an action potential in the receiving cell that will result in a continuation of the message.

There are about one quadrillion synapses in the brain of a 3-year-old child (Chanhangeux and Ricoeur 2000), which will decline with age, while a typical neuron has 1000–10,000 synapses which might receive information from 1000 nerve cells. Despite this phenomenally huge number, all the nerve cells together account for only about half the volume of the brain. The remainder of the brain's volume is comprised of a variety of support cells, the most important of which are the *glial* cells (glial originated from the Latin word meaning glue). It is estimated that there are about one trillion glial cells within the CNS. The glia occupy essentially all of the space in the CNS which is not filled by neurons, and there are several different types of these cells in the brain, with each having a special role.

Astrocytes (Astroglia) are the largest of the glial cells, named for their starlike shape. Some astrocytes are wrapped around the dendritic and somatic portions of the neuron, while others form a layer around blood vessels. It has been hypothesized that astrocytes supply neurons with nutrients from the blood, but astrocytes may also be important in transporting ions from the brain to the blood. As some astrocytes surround the synapses, they limit the dispersion of neurotransmitters which are released into the synaptic cleft and prevent cross-talk among neurons which may be performing different functions. Certain astrocytes also act as phagocytes (scavenger cells that clean up the debris of neurons killed by head injury, infection, or heart stroke). A significant and interesting property of neurons is that they do not undergo cell division; hence, if a neuron is fatally damaged, it is lost forever.

Another type of glial cells is oligodendrocytes, or oligodendroglia, which provide support for axons and coat axons in the brain and spinal cord with myelin. Myelin serves to insulate axons from one another and is made of lipid and protein. It is produced by the oligodendrocytes in the form of a tubular sheath that surrounds the axon (Fig. 4.1). The sheath is not continuous, but consists of a series of segments, with a small portion of each axon left uncoated between segments. This uncoated segmental end is called the node of Ranvier. Myelin isolates axons from each other and helps conserve the neuron's metabolic energy. A third type of glial cells, the microglia, consists of extremely small cells which remain dormant until damage occurs in the CNS. They are then transformed into phagocytes, which, like certain of the astrocytes, move throughout the brain to clean up the debris left by dead neurons.

4.2 Development of the Central Nervous System (CNS)

A single fertilized egg gives rise to all cells in human body. In the second week after conception, and following implantation in the uterine wall, the blastocyst differentiates into three distinct cell layers, the endoderm, the mesoderm, and the ectoderm, which serve as the body's tissues. Ectoderm is the origin of both the skin and the nervous system. By the third week of conception, the first perceptible sign of body formation appears, specifically a pear-shaped neural plate which arises from the dorsal ectoderm. All the cells that become neurons develop from this neural plate. Shortly after formation, a fold appears in the neural plate which joins together in forming the neural tube, which contains approximately 125,000 cells and is the precursor of the CNS. Once the neural tube is fully formed, the mass of cells is officially called the embryo (Gilbert 2013). A number of anomalies result if the neural tube does not close properly, such as anencephaly, where the forebrain fails to develop correctly, because the anterior portion of the neural tube does not close. Increased folic acid in the diet can preempt such malformation.

During the fourth month after conception, once the neural tube is closed, the nerve cells begin to proliferate rapidly. Based on an average gestation period of 270 days, neurons are produced at the rate of 250,000 per minute during prenatal development. With few exceptions, neurons in the human nervous system stop forming around the time of birth, which means that the 100 billion neurons which comprise the adult's brain are actually present in the brain of the newborn. The period of proliferation does vary for different populations of cells, however. But, in general, the proliferation of larger neurons which tend to serve long-distance functions within the CNS precedes the proliferation of smaller cells which are destined primarily for "local circuits" within the brain. Before they differentiate into their final form, nerve cells are referred to as neuroblasts. When they have finished dividing, neuroblasts migrate to their ultimate location within the brain. The radial glia, one of the glial cells, play an important function in neural migration. Present only during prenatal development and disappearing shortly before birth, they extend from the inner to outer surfaces of the emerging nervous system. The radial glia appear to act as guide wires on which the migrating neuroblasts move slowly to their destinations. Some subsequently forming neuroblasts migrate in a different manner, being attracted to the surfaces of the nerve cells and migrating along the axons of earlier formed neurons. Following migration to its final location, a neuroblast must orient itself properly and associate with other similar cells. The process called aggregation is important for the development of functional units of the CNS, for example, layers of the cortex. Neural cell adhesion molecules, on the neural surface, promote the aggregation of nerve cells, destined to have similar functions (Alberts et al. 2008).

To begin with, neuroblasts resemble mature nerve cells no more than what they do of the other cells in the body. Once when they arrive at the destination, neuroblasts begin to acquire the distinctive appearance of the characteristic of the neurons in its specific location in the CNS. During this process of differentiation, each neuron will develop its own specific dendritic pattern, axonal structure, and distinctive neurotransmitter properties.

When nerve cells begin to differentiate, they must also make appropriate connections with other nerve cells. As noted earlier, connections between neurons occur at the synapse. Synaptogenesis, or the formation of synapses, begins during the prenatal development, but it is not confined to this period. Indeed, the formation of most synapses in the human neocortex occurs after the birth. This process actually accounts for the largest change in brain cells between birth and adulthood. A variety of environmental factors, notably nutrition, can alter postnatal synaptogenesis. This is where optimum nutrition has a crucial role to play in human development.

Despite a hundred and odd billion neurons which are present at birth, far more nerve cells are produced than are actually found in the newborn baby's brain. Rather surprisingly, extensive cell death is a crucial phase of brain development. Based on the location of the brain, about 15–85% of neurons die during fetal development. One possible explanation for the death of some neurons versus survival of others is competition for the life-maintaining linkages they must make with their proper synaptic targets. That is, neurons which do not make appropriate synaptic linkages would most likely die. As with synaptogenesis, the death of the nerve cells continues after birth. In fact, it has been suggested that behavioral maturity results not from synaptic formations which occur after birth instead from the elimination of excess connections and the increasing efficiency of those connections which remain in the human brain (Bloom et al. 2006).

With all the neurons in place, a great deal of growth still ahead, at birth the human brain weighs between 300 g and 350 g. In fact, the infant's brain grows quite rapidly during early postnatal development, doubles in weight by 1 year, reaching 80% of the adult brain weight by the age 4. If not the nerve cells, what else grows? There are three to four areas of growth which appear to continue and are critical in brain development. The first area is the glial cells that proliferate rapidly at around 30 weeks' gestation and continue to develop throughout life. It is shortly after birth that the most intense growth in glial cells occurs in humans and many other mammalian species. The second area is the development of the myelin sheath around the axons of the nerve cells. Myelination greatly hastens the rate at which axons conduct messages, as well as keeps them from interacting with proximal neurons which may be programmed for different functions. In humans, rapid myelination begins shortly after birth and continues until the child is of age four. The first nerve tracts in the CNS to become myelinated are in the spinal cord, followed by the hindbrain, midbrain, and forebrain in this sequence. Within the cerebral cortex, sensory neurons are myelinated before motor neurons, and as a result, sensory function precedes motor function. The third area of growth is in the aforementioned synaptogenesis, which as already indicated represents the largest increase in brain cells between birth and maturity.

Some recent research suggests that certain nerve cells may, in fact, develop postnatally. For instance, there are few areas around the brain ventricles in which the proliferation of neurons remains evident after birth. Small neurons which are derived from these areas may be added to portions of the brain such as the hippocampus throughout the first few years of postnatal life (Rosenzweig and Leiman 1989).

4.3 The Development of the Brain

4.3.1 Effect of Nutrients

Both macro- and micronutrients affect brain development. The nutrients may be categorized into three classes as follows:

Macro
Secondary
Micro (also known as trace elements)

Some examples:

Macronutrients: nitrogen (N), phosphorus (P), potassium (K), carbon (C), hydrogen (H), and oxygen (O_2)
Secondary nutrients: calcium (Ca), magnesium (Mg), and sulfur (S)
Micronutrients or trace elements: zinc (Zn), manganese (Mn), copper (Cu), iron (Fe), molybdenum, (Mo), iodine (I), etc.

It is very interesting to note that from an evolutionary point of view, the nutrients required by plants or which the plants contain are also required by humans and animals. As an extension of this thought, it could be argued that without plants, which supply these nutrients from their cells to humans and animals, no life on planet earth is possible. Thus, plants form the link between earth and life, and for the sustenance of plants, a good soil, with abundant fertility (denoted by the abundant presence of these nutrients), is a must.

The discussion on the effect of nutrients will focus on both the macro- and micronutrients, which affect behavior. For the moment, how their deficiency will impact brain lipids is of interest, and for much of our knowledge, conclusions are derived from animal studies. In animal experiments, it has been found that suckling rats with restricted food intake have lower brain weight, protein, and lipid contents, with phospholipids and cholesterol reduced by 80%, relative to the normally fed littermates (Perry et al. 1986). Vitamin (thiamine, for instance) deficiency leads to impaired deposition of cerebrosides, phospholipids, and cholesterol (Reddy and Ramakrishnan 1982), while niacin deficiency impairs cerebroside content and myelination (Nakashima and Suzue 1982), and peroxidase deficiency reduces levels of myelin lipids and polyunsaturated fatty acids in the cerebellum. Folate-deficient mothers (human) appear to have a greater likelihood of delivering infants who display malformations of the CNS, including microcephaly (Oakley et al. 1994).

Micronutrients (trace elements) exert a greater role in brain functioning than on its development, although Zn (zinc) and Cu (copper) deficiencies are particularly damaging to the maturation of the brain (Odutaga 1982). Na (sodium) and K (potassium) are necessary for electrical and antioxidant activity, fluid balance, and synaptic communication. Se (selenium) facilitates antioxidant activity, while Ca (calcium), Co (cobalt), Cu (copper), Fe (iron), Mg (magnesium), Mn (manganese), Mo (molybdenum), and Zn (zinc) are all essential for brain function (Levander and Burk 1992).

4.3.2 Effect of Lipids and Fatty Acids

The prenatal period is critical for brain growth, as the number of neurons and the myelination process are being established. The first postnatal year is also important, as the infant brain continues to grow by approximately 120%. In fact, some 60% of the infant's total energy intake during the first year after birth is used by the brain, in constructing neuronal membranes and depositing myelin (Tacconi et al. 1997). Most of this energy intake comes from dietary fat, with lipids accounting for over half the dry weight of the brain. A vast number of compounds fall under the definition of lipids, which are variably distributed in the gray and white matter of the brain, as well as in neurons and glia. The lipids found most in the brain are cholesterol, phospholipids, cerebrosides, gangliosides, sulfatides, and an array of fatty acids.

Cholesterol is a key component of the neuronal plasma membrane, indeed, of all plasma membranes and regulates and maintains the internal milieu of the nerve cell. The basic structural unit of the membrane is the plasma membrane lipid bilayer, which provides a complex chemical environment for the protein molecules which mediate the cell function (Mason et al. 1997). During intrauterine growth, the fetus receives no cholesterol of dietary origin, but synthesizes its own using glucose and fatty acids. At birth, most of the plasma cholesterol comes from high-density lipoprotein. In the first weeks after birth, the concentration of plasma total and lipoprotein cholesterol rise steeply until they reach levels that are maintained until adulthood (Kwiterovich 1986). Alterations in neuronal plasma membrane cholesterol content seem to be important in modulating the activity of neurotransmitter receptors; hence, cholesterol also appears to be involved in the regulation of brain function (Dietschy and Turley 2004).

Phospholipids, also crucial in building bilayers of the different brain cell membranes, play an important role as secondary messengers and signal mediators (Shukla and Halenda 1991). Apart from the necessity of gangliosides to the development of ganglion, the specific roles of lipids such as cerebrosides and sulfatides have been less clearly established, although their abundance in nerve axons suggests their importance in nerve impulse conduction (Hannun and Bell 1989).

Fatty acids are normally present as esters in phospholipids, and their composition has an impact on the fluidity of membranes (Spector and Yorek 1985). When released after stimulation, some of them act as secondary messengers or may be precursors of eicosanoids, hormonelike chemical regulators that influence blood vessel dilation and constriction (Insel et al. 2004). The fatty acid profile of the brain is unique in that it contains large amounts of essential fatty acids. These fatty acids become polyunsaturated acids, whose availability is a limiting factor in brain development (Tacconi et al. 1997).

4.3.3 Effect of Polyunsaturated Fatty Acids and Early Behavior

The long-chain polyunsaturated fatty acids (LCPUFAs) which are located on the cell membrane phospholipids serve as important structural components of the brain. Investigations on LCPUFAs represent one of the most rapidly growing areas of research in the study of early human development and functioning. Two major brain LCPUFAs are docosahexaenoic acid (22:6n–3) and arachidonic acid (20:4n–6). The abbreviations in parentheses reflect the standard nomenclature for fatty acids, with the first number representing the length of the carbon chain, the number before the letter n representing the number of double bonds, and the number after the dash referring to the number of carbon atoms from the methyl end of the molecule to the first double bond. Because they are prevalent in green plants, algae, and phytoplankton on which fish feed, fish oils are a rich source of docosahexaenoic acid (DHA), while egg lipids can provide both DHA and arachidonic acid (ARA). Two fatty acids, namely linoleic acid (18:2n–6 also known as an omega-6 fatty acid) and alpha-linolenic acid (18:3n–3 also known as omega-3 fatty acid), are precursors of the LCPUFAs and must be obtained from the diet because they cannot be synthesized (see Table 4.1).

For the above-cited reason, they are termed essential fatty acids (EFAs). Once provided through the diet, the CNS and liver have enzymes which can convert them into longer-chain polyunsaturated fatty acids. Vegetable oils are a rich source of both linoleic and alpha-linolenic acids.

Omega-6 to omega-3 acid ratio was approximately 1:1 in early humans, but the typical western diet today has a ratio varying from 10:1 to 25:1 due to enhanced use of vegetable oils rich in linoleic acid as well as reduced fish consumption. Increasing research evidence shows that enhanced omega-3 fatty acid content in diet leads to better health benefits (Haag 2003). High DHA in the retina – in some ways an extension of the brain – suggests better visual acuity. Non-myelin membranes of the CNS contain proportionally high amounts of DHA and ARA (Innis 1997). Hence, a number of organ functions are likely to be affected by its uptake. In the brain, DHA is most abundant in membranes which are associated with synaptic function and is accumulated in the CNS late in gestation and early in postnatal life. Both animal studies and infant feeding with formulas containing DHA have shown better brain function.

Table 4.1 The array of polyunsaturated fatty acids

Main fatty acids	Fatty acid carbon number
Ω–3 s	
Hexadecatrienoic acid (HTA)	16:3n–3
↑ Linolenic acid (ALA)	18:3n–3
Stearidonic acid (SDA)	18:4n–3
Eicosatrienoic acid (ETE)	20:3n–3
Eicosatetraenoic acid (ETA)	20:4n–3
Eicosapentaenoic acid (EPA)	20:5n–3
Docosapentaenoic acid (DPA)	22:5n–3
Docosahexaenoic acid (DHA)	22:6n–3
Tetracosapentaenoic acid	24:5n–3
Tetracosahexaenoic acid	24:6n–3
Ω–6 s	
Linoleic acid	18:2n–6
y-Linolenic (GLA)	18:3n–6
Eicosadienoic acid	20:2n–6
Dihomo-y-linolenic (DGLA)	20:3n–6
Arachidonic acid (ARA)	20:4n–6
Eicosapentaenoic acid (mammal)	20:5n–6
Docosadienoic acid	22:2n–6
Adrenic acid	22:4n–6
Docosapentaenoic acid (Osbond)	22:5n–6
Tetracosatetraenoic acid	24:4n–6
Ω–9 s	
Mead acid	20:3n–9

In research with rhesus monkeys, it was observed that a diet deficient in omega-3 fatty acids led to changes in both photoreception and cortical functions related to the visual system (Connor et al. 1992). This may explain the longer time that the monkeys spend in looking at test stimuli, a behavior which actually suggests lower cognitive competence (Reisbick et al. 1997). Experiments with rats have shown that diets deficient in omega-3 fatty acids produce cognitive deficits (Catalan et al. 2002), abnormal retinal functioning (Bazan 1990), and heightened motor activity, as indicated by increased exploration of novel environments (Enslen et al. 1991). However, a rat that spends more time to explore may not spend as much time trying to learn a new task, and the confound between visual acuity and learning may also be problematic (Benton 1997). Regardless of the learning task, rats given diets containing omega-3 or omega-6 fatty acids versus diets deficient in essential fatty acids (EFAs) have been more successful in performing on traditional mazes, water mazes, and shock avoidance tasks. Although the impact on their learning cannot always be distinguished from other mental abilities, the consistency of these findings strongly suggests that the fatty acid content of the diet influences learning and memory (Benton 1997).

4.3.4 Interrelationship Between Infant Development and Essential Fatty Acids (EFAs)

Most infants do not obtain specific PUFAs in their diet if they are being fed with a factory-manufactured infant formula. To explain, breast milk contains DHA, but standard formulas in USA traditionally do not. In the developing world, especially in rural areas this is not the case, because, for more than 6 months after birth, the infant's principal diet is breast milk. In western countries due to increasing attention on physical appearance, women after child birth dispense with breast-feeding. In fact, the whole practice is against nature. In the mid-1990s, the UK and Canada began manufacturing omega-3 formulas available to consumers for infant use. Recent changes in the composition of commercial formulas illustrates how policy makers weigh scientific evidence before making their decisions yet may still be influenced by lobbyists. The European Society for Paediatric Gastroenterology and Nutrition in 1991 and British Nutrition Foundation in 1992 recommended not only that omega-3s be present but that DHA and ARA be added to infant formulas. In response, in 1995 the Food and Drug Administration (FDA) of the US commissioned the American Society for Nutritional Sciences to appoint an interdisciplinary team of nutrition experts to issue of supplementing infant formulas to make recommendations regarding their implementation. The findings published in the *Journal of Nutrition* in 1998 led them to suggest minimum and maximum levels for omega-3s, but they refrained from giving manufacturers the go ahead to add DHA to infant formula. Given the relevance of this issue, to the role of nutrition in brain and behavior connections, one must consider the following evidence the panel suggested. The prime objective of the expert panel was to look into the role of either macro- or micronutrients in infant formula foods. They used six types of supportive evidence to determine what the minimum levels should be for each conceivable nutrient, namely, direct experimental evidence, metabolic balance studies, clinical observations of deficiency, extrapolation from experimental evidence to older age groups, theoretically based calculations, and its analogy to breast-fed infants. The last criterion is particularly meaningful, because the utility of formula has always been evaluated by comparing its effects on infant growth and development to how babies fare if they are fed with breast milk. Indeed, breast milk is considered "the gold standard" to which all formulas should be compared.

4.3.5 The Eternal Comparison: Breast Milk vs. Formulas

This is an eternal question that has baffled human mind. But, if one meticulously examines the evolutionary process, whether human or animal, the mother's milk is the basic source of energy for the infant. Legendary evidence from the Hindu mythology shows that Yashoda, Lord Krishna's mother, when skipped a cycle of breast-feeding, the milk oozed out from the full breasts. This also can be seen among cows and buffaloes. When the calves are prevented from feeding, the milk will ooze from the teats.

A vastly prevalent notion is that breast-fed infants are smarter than formula fed ones. Hoefer and Hardy (1929) reported that children of 7–13 years, breast-fed, had higher IQs than those formula-fed, or weaned by 3 months after birth. In the last decades of the twentieth century, investigators again began to look for differences in intelligence between breast-fed and formula-fed children. In the western world, breast-feeding is associated with higher socioeconomic status (SES), while in developing and poorer countries, it is a way of life. Higher SES is a marker for more involved parenting, which usually translates into more attention being paid to the infant – whether through reading, playing, or all round caregiving. This, in turn, has been shown to facilitate cognitive development in childhood (Bradley and Corwyn 2002). To address this important confound, efforts have to be made to try to control for SES while evaluating the impact of breast-feeding on childhood intelligence. A meta-analysis of such efforts concluded that children who were breast-fed have IQ scores from 2–3 points higher than children formula-fed (Anderson et al. 1999). Another study indicates that the duration of breast-feeding would, as well, have an influence on the IQ. This would display a dose-response effect on adult intelligence (Mortensen et al. 2002). More recently, a large-scale study that randomized partici-pants (who had already decided to breast-feed) to a breast-feeding promotion inter-vention reported a difference of 5.9 score in IQ favoring the 6.5-year-old children, whose mothers were in the experimental group (Kramer et al. 2008). However, ongoing research suggests that other family factors, like the maternal IQ and paren-tal education, may better account for higher childhood IQ scores than does breast-feeding in and of itself (Colen and Ramey 2014).

The positive aspect of breast-feeding on infant IQ shows that while breast milk contained DHA, its absence in the formula feed made the difference. DHA, as dis-cussed earlier, is one of the most important fatty acids among the LCPUFAs. Though experimental results on this issue are often confusing and contradictory, one needs to look at results of investigation where LCPUFA supplementation and infant behavior is studied. In the mid-1990s, the Expert Panel reviewed some 30 investiga-tions which specifically focused on LCPUFAs and infant behavior. Since infants are preverbal, the tasks used to determine their abilities are of a different form than what one would use if one were testing intelligence, in, say, a 10-year-old child. Since the retina is laden with DHA, a number of investigators have used what amounts to a preferential looking technique in order to test the infant's acuity. Although some of the investigators have used this approach and found LCPUFA formula to result in better visual acuity, than a standard formula, particularly in preterm infants, a num-ber of other sound investigations have not. However, in a recent meta-analysis of 19 investigations, it was found that LCPUFA supplementation of infant formula improved infant visual acuity up to 12 months, albeit as tested using visual-evoked potential – a method more objective than the visual preference technique (Qawasmi et al. 2012).

Testing the real IQ in a preverbal child is quite tenuous. Infant test is one of the approaches for this (Fagan and Shephers 1987). Also relying on the infant's ability to look, the infant is first given the opportunity to habituate to one stimulus. When a new stimulus is paired with the first, the baby is expected to spend more time looking at the novel stimulus. A retrospective study with infants whose formula contained either omega-3 or omega-6 fatty acids revealed that there was no difference in novelty preference with this test (Innis et al. 1996).

The most popular assessment tool for testing infant mental performance, in contrast to all the above-cited examples, is the "Scales of Infant Development," in which sensory orientation, fine and gross motor skills, imitative ability, and early receptive and expressive language are tested (Bayley (1993)). The test comes closest to what psychologists would term measuring infant "intelligence," although terming it "cognitive development" would be a more appropriate description.

It is important to suggest that there is no irrefutable positive scientific evidence on DHA or LCPUFA on infant behavior, though implied benefit does exist. The Expert Panel recognized the weakness of earlier studies, including small sample size, variable demography, short-term follow-ups, etc., as a result of which, it did not recommend the addition of DHA or ARA to infant formulas at that time, although minimum levels for omega-3 and omega-6 fatty acids were approved. Aware that this was a rather recent area of research, the panel endorsed the continuation of the basic science and clinical studies to further examine the role of LCPUFAs in infant development and recommended a reappraisal of newer data by 2003. In 1997, a second Expert Panel began examining the scientific evidence which pertained to nutrient requirements for preterm infant formulas. In their 2002 report, also published as a special issue of the *Journal of Nutrition*, the panel did make recommendations for a minimum and maximum ARA:DHA ratios but reiterated their non-endorsement of ARA:DHA levels in formula for full-term infants (Klein 2002).

The extent of corporate pressure can be gauged by the fact that FDA (Food and Drug Administration of the USA) gave approval in 2001 for companies to market such formulas, as discussed above, for full-term infants instead of waiting until a subsequent Expert Panel assessed more recent data. Nonetheless, some research suggests that newer measures of infant cognitive development may reveal more robust effects. For example, using the 1993 version of the Bayley scales, it was shown that supplementation of formula with DHA or DHA + ARA did improve mental scores (Birch et al. 2000). In fact, scores were higher for DHA + ARA group, followed by DHA group, which was followed by the control group. However, the bulk of recent research has failed to show significant differences between LCPUFA-supplemented and control groups on the Bayley scales, as summarized in a recent meta-analysis of 12 randomized controlled clinical trials with full-term and preterm infants (Qawasmi et al. 2012). The International Formula Council (IFC) takes the position that DHA and ARA are the "building blocks" of infant eye tissue and the developing brain, advancing a neural development argument.

4.3.6   Adult Cognitive Capability Linked to LCOPUFA

Recent research has focused on brain function in the elderly and most notably the possible effects of LCPUFAs in preventing cognitive decline. It is estimated that between 2010 and 2050, the total number of individuals over 60 years of age, with care needs worldwide, will have increased from 29 percent to 45 percent or to nearly 277 million, and about half of those who will require personal care will suffer from dementia (Prince et al. 2013). Apart from Alzheimer's disease (AD), with increasing age, many other individuals experience mild cognitive impairment (MCI) or cognitive impairment no dementia (CIND).

Lower levels of omega-3 fatty acids, such as DHA, in serum, in erythrocyte membranes, and in the brain tissue (taken post mortem) of patients with AD and CIND (Whalley et al. 2008) suggest that diet containing these important constituents have an important role in cognitive performance in the elderly. Evidence also suggests that dietary intake of fatty acids by eating fish is associated with better cognitive outcomes in the elderly, though some studies contradict this. Nevertheless, some prospective studies with the elderly indicate that a higher intake of omega-3 fatty acids may be protective against cognitive disabilities, as well as reducing their rate of cognitive decline (Devore et al. 2009).

There is, yet, another measure called Mini-Mental State Examination (MMSE) used to screen cognitive impairment and dementia. A clinician asks the individual 30 questions which cover memory, following directions, counting/spelling backward, and orientation (e.g., "Where are you"?). A score of 23 or less than 23 indicates cognitive impairment. While not perfect, the MMSE has frequently been used in combination with other tests to assess the effect of fatty acids on cognitive performance (Yurko-Mauro et al. 2010).

4.3.7   Cholesterol-Linked Adult Behavior

Although the link between controlling serum cholesterol and coronary heart disease (CHD) is more or less established, some investigations showed that while deaths due to CHD did show a decline among treated patients, the number of deaths due to non-cardiovascular causes, such as suicides, homicides violence, and accidents, all behavioral phenomena, actually seemed to increase. In this connection, it must be stated that statin use may decrease the risk of depression, a behavioral aberration (Otte et al. 2012), though results may differ by sex, with fewer depression cases in women compared to men (Feng et al. 2010).

4.3.8 The Cholesterol: Antisocial Behavior Link

The relationship between cholesterol and non-illness-related deaths remain controversial, and there is a body of intriguing investigation results which suggest a link between lower serum cholesterol (below 160–180 mg/dl) to psychiatric and behavioral manifestations of affective disorders and violence. For example, individuals with antisocial personality disorder, whether of a psychopathic or sociopathic nature, have been shown to have lower serum cholesterol (Repo-Tiihonen et al. 2002). Also, low-cholesterol concentrations have also been observed in prisoners, homicidal offenders, patients hospitalized for violence, and those who attempt suicide (Kaplan et al. 1997). Low cholesterol may directly influence mood and suicidal behavior but, perhaps, as likely, mood and medication, via their influence on eating and exercise, which may serve to reduce cholesterol levels (Zhang 2011).

Apart from suicide, the association between plasma cholesterol and aggressive behavior under laboratory conditions has been investigated using Asian monkeys (Kaplan et al. 1997). The cynomolgus macaque is closely related to the rhesus monkey but about half the size. In a series of carefully controlled investigations, the monkeys were raised in social groupings, were fed diets that were either high or low in saturated fat and cholesterol, and were subjected to routine plasma lipid sampling, as well as behavioral patterns (Kaplan et al. 1997). Depending on the specific study, monkeys in the low-fat conditions exhibited more aggressive behavior involving physical contact, spent more time alone if they consumed the low-cholesterol diet and sought less body contact with their peers (Kaplan et al. 1994). Clearly, then, a low-fat or low-cholesterol diet for these monkeys resulted in a variety of antisocial behavioral patterns.

4.3.9 How Does Cholesterol Affect Cognition?

Based on animal-centered investigations, it has been shown that dietary manipulations would modify human behavior by changing brain cholesterol levels and the fluidity of neuronal membranes. Nevertheless, animal studies are inconclusive, as both elevated serum cholesterol and decreased serum cholesterol have been associated with enhanced water maze learning in rats and increased eye-blink conditioning in rabbits (Schreurs 2010). With humans, almost no research has been conducted on the role of cholesterol and learning ability, although some other aspects of cognitive functioning have been explored, such as memory. For instance, a study with non-demented, though very elderly individuals, found high total cholesterol to be associated with better memory function (West et al. 2008). Research on cholesterol and its effects on the elderly has been of particular interest in behavioral research. Using MMSE as an outcome measure, a number of investigators have found higher levels of cholesterol to be optimal for elderly individuals, whether it is of the high-density lipoprotein (HDL) form – the proverbial "good" cholesterol.

Contrary results, showing higher dementia with elevated total cholesterol among the elderly, have also been shown (Solomon et al. 2009).

One area of research interest is the effect of statins on cholesterol vis-à-vis cognition. The consensus is that statins improve cognition (Carlsson et al. 2008). In contrast to the fairly large number of investigations with adults of various ages, on this above aspect, investigations with children, which are but rare, show that there is no perceptible effect of cholesterol either on cognition or academic performance (Perry et al. 2009). Beyond childhood, however, the current assessment appears to be that higher levels of cholesterol are most detrimental in middle-age adults and most beneficial in the elderly (Schreurs 2010).

4.3.10   Cholesterol-Serotonin Link

The question, by what mechanism does dietary cholesterol affect brain function, or in simple terms, human behavior, is an important one. Taken together, the foregoing review has built up a rather intriguing, though tentative, case for what has been termed the cholesterol-serotonin hypothesis. Almost a quarter century ago, Engelberg (1992) proposed that reduced serum cholesterol might be accompanied by decreased serotonin precursors, as well as changes in the function of serotonin receptors and transporters which could cause an increase in suicidal thoughts. Serotonin and serotonergic activity have been shown to play an important role in mood disorders. Results of Kaplan and colleagues, (Kaplan et al. 1997), discussed earlier, show that in cynomolgus monkeys, reduced serum cholesterol leads to a decrease in neurotransmitter serotonin, leading to aggressive behavior. This hypothesis presumes the following three associations:

An inverse relationship between plasma cholesterol and aggressive behavior
A positive association between serum cholesterol and central serotonergic activity
A link between reduced central serotonergic activity and increased aggressive or
    violent behavior

From their investigations, Kaplan and colleagues suggested that the results may be useful in making sense of the epidemiological associations found between low serum cholesterol and increased incidence of violent deaths and even speculated that early in history, a cholesterol-serotonin-behavior linkage may have served as a mechanism to increase competitive behavior for high-fat foods (Bunn and Ezzo 1993). Observational studies with humans have shown a trend of lower serotonin in those with lower cholesterol. Some research with humans has also demonstrated that naturally low and experimentally lowered serotonin can attenuate violent behavior (Kruesi et al. 1990) with some suggesting an evolutionary value (Wallner and Machatschke 2009). The cholesterol-serotonin hypothesis is an intriguing starting point to help us better understand the paradox of lowering cholesterol being shown as ineffective in reducing mortality. Probably, researchers over the coming decades will most likely continue to investigate this hypothesis to further delineate the role of cholesterol and other lipids in facilitating brain development and function.

References

Alberts, B., Johnson, A., Lewis, J., Raff, M., Roberts, K., & Walter, P. (2008). *Molecular biology of the cell* (5th ed.). New York: Garland Science.

Anderson, J. W., Johnstone, B. M., & Remley, D. T. (1999). Breast-feeding and cognitive development: A meta-analysis. *American Journal of Clinical Nutrition, 70*, 525–535.

Azevedo, F. A. C., Carvalho, L. R. B., Grinberg, L. T., Farfel, J. M., Ferretti, R. E. L., Leite, R. E. P., Filho, W. J., Lent, R., & Herculano-Houzel, S. (2009). Equal numbers of neuronal and nonneuronal cells make the human brain as isometrically scaled-up primate brain. *Journal of Comparative Neurology, 513*, 541.

Bayley, N. (1993). *Bayley scales of infant development* (2nd ed.). San Antonio: Psychological Corporation.

Bazan, N. G. (1990). Supply of n-3 polyunsaturated fatty acids and their significance in the central nervous system. In R. J. Wurtman & J. J. Wurtman (Eds.), *Nutrition and the brain* (Vol. 8, pp. 1–24). New York: Raven Press.

Benten, D. (1997). Dietary fat and cognitive functioning. In M. Hillbrand & T. Spitz (Eds.), *Lipids, health, and behavior* (pp. 227–243). Washington, DC: American Psychological Association.

Birch, E. E., Garfield, S., Hoffman, D. R., Uauy, R., & Birch, D. G. (2000). A randomized control trial of early dietary supply of long chain polyunsaturated fatty acids and mental development in term infants. *Developmental Medicine and Child Neurology, 42*, 174–181.

Bloom, R. E., Nelson, C. A., & Lazerson, A. (2006). *Brain, mind and behavior* (3rd ed.). New York: W.H. Freeman.

Bradley, R. H., & Corwyn, R. F. (2002). Socioeconomic status and child development. *Annual Review of Psychology, 53*, 371–399.

Bunn, H. T., & Ezzo, J. A. (1993). Hunting and scavenging by Plio-Pleistocene hominids: Nutritional constraints, archaeological patterns, and behavioural implications. *Journal of Archaelogical Science, 20*, 365–398.

Carlsson, C. M., Gleason, C. E., Hess, T. M., Moreland, K. A., & Blazel, H. L. (2008). Effects of simvastatin on cerebrospinal fluid biomarkers and cognition in middle -aged adults at risk for Alzheimer's disease. *Journal of Alzheimer's Disease, 13*, 187–192.

Catalan, J., Toru, M., Slotnick, B., Murthy, M., Grener, R. S., & Salem, N., Jr. (2002). Cognitive deficits in docosahexaenoic acid-deficient rats. *Behavioral Neuroscience, 116*, 1022–1031.

Chanhangeux, J. P., & Ricoeur, P. (2000). *What makes us think?* Princeton: Princeton University Press.

Colen, C. G., & Ramey, D. M. (2014). Is breast truly best? Estimating the effects of breastfeeding on long-term child health and wellbeing in the United States using sibling comparisons. *Social Science & Medicine, 109*, 55–65. https://doi.org/10.1016/j.socscimed.2014.01.027.

Connor, W. E., Neuringer, M., & Reisbick, S. (1992). Essential fatty acids. The importance of n-3 fatty acids in the retina and brain. *Nutrition Review, 50*, 21–29.

Dangour, A. D., & Allen, E. (2013). Do omega-3 fat boost brain function in adults? Are we any closer to an answer? *American Journal of Clinical Nutrition, 97*, 909–910.

Devore, E. E., Grodstein, F., van Rooij, F. J. A., Hofman, A., & Rosner, B. (2009). Dietary intake of fish and omega-3 fatty acids in relation to long-term dementia risk. *American Journal of Clinical Nutrition, 90*, 170–176.

Dietschy, J. M., & Turley, S. D. (2004). Cholesterol metabolism in the central nervous system during early development and in the mature animal. *Journal of Lipid Research, 45*, 1375–1397.

Engelberg, H. (1992). Low serum cholesterol and suicide. *The Lancet, 339*, 727–729.

Enslen, M., Milon, H., & Malone, A. (1991). Effect of low intake of n-3 fatty acvids during development on brain phospholipids fatty acid composition and exploratory behavior in rats. *Lipids, 26*, 203–208.

Fagan, J. F., & Shephers, P. A. (1987). *The Fagan test of infant intelligence*. Cleveland: Infantest Corporation.

Feng, L., Yap, K. B., Kua, E. H., & Ng, T. P. (2010). Statin use and depressive symptoms in a prospective study of community-living older persons. *Pharmacoepidemiology and Drug Safety, 19*, 942–948.

Gilbert, S. F. (2013). *Developmental biology* (10th ed.). Sunderland: Sinauer Associates.

Haag, M. (2003). Essential fatty acids and the brain. *Canadian Journal of Psychiatry, 48*, 195–203.

Hannun, Y. A., & Bell, R. M. (1989). Functions of sphingolipids and sphingolipid breakdown products in cellular regulation. *Science, 243*, 500–507.

Hoefer, C., & Hardy, M. C. (1929). Later development of breast fed and artificially fed infants: Comparison of physical and mental growth. *The Lancet, 92*(8), 615–619.

Innis, S. M. (1997). Polyunsaturated fatty acid nutrition in infants born at term. In J. Dobbing (Ed.), *Developing brain and behavior: The role of lipids in infants formula*. San Diego: Academic.

Innis, S. M., Nelson, C. M., Lwanga, D., Rioux, F. M., & Waslen, P. (1996). Feeding formula without arachidonic acid and docosahexaenoic acid has no effect on preferential looking acuity or recognition memory in healthy full term infants at 9 months of age. *American Journal of Clinical Nutrition, 64*, 40–46.

Insel, P., Turner, R. E., & Ross, D. (2004). *Nutrition* (2nd ed.). Sidbury: Jones and Bartlett Publishers.

Kaplan, J. R., Shively, C. A., Botchin, M. B., Morgan, T. M., Howell, S. M., Manuck, S. B., Muldoon, M. F., & Mann, J. J. (1994). Demonstration of an association among dietary cholesterol, central serotonergic activity, and social behavior in monkeys. *Psychosomatic Medicine, 56*, 479–484.

Kaplan, J. R., Manuck, S. B., Fontenot, M. B., Muldoon, M. F., Shively, C. A., & Mann, J. J. (1997). The cholesterol-serotonin hypothesis: Interrelationships among dietary lipids, central serotonergic activity, and social behavior in monkeys. In M. Hillbrand & R. T. Spitz (Eds.), *Lipids, health, and behavior*. Washington, DC: American Psychological Association.

Klein, C. J. (Ed.). (2002). Nutrient requirements for preterm infant formulas. *The Journal of Nutrition, 132*(6), Suppl.1. 1395S-577S.

Kramer, M. S., Aboud, F., Mironova, E., Vanilovich, I., Plat, R. W., & Matusch, L. (2008). Breastfeeding and child cognitive development: New evidence from a large randomized trial. *Archives of General Psychiatry, 65*, 578–584.

Kruesi, M. J. P., Rapport, J. L., Hamburger, S., Hibbs, E., & Potter, W. Z. (1990). Cerebrospinal fluid monoamine metabolites, aggression, and impulsivity in disruptive behavior disorders of children and adolescents. *Archives of General Psychiatry, 47*, 419–442.

Kwiterovich, P. O. (1986). Biochemical, clinical, epidemiologic, genetic, and pathologic data in the pediatric age group relevant to the cholesterol hypothesis. *Pediatrics, 78*, 349–362.

Levander, O. A., & Burk, R. F. (1992). Selenium. In M. L. Brown (Ed.), *Present knowledge in nutrition* (6th ed., pp. 268–273). Washington, DC: International Life Science Institute-Nutrition Foundation.

Mason, R. P., Rubin, R. T., Mason, P. E., & Tulenko, T. N. (1997). Molecular mechanisms underlying the effects of cholesterol on neuronal cell membrane function and drug-membrane interactions. In M. Hillbrand & R. T. Spitz (Eds.), *Lipids, health, and behavior*. Washington, DC: American Psychological Association.

Mortensen, E. L., Michaelsen, K. F., Sanders, S. A., & Reinisch, J. M. (2002). The association between duration of breastfeeding and adult intelligence. *Journal of the American Medical Association, 287*(18), 2365–2371.

Nakashima, Y., & Suzue, R. (1982). Effect of nicotinic acid on myelin lipids in brain of the developing rat. *Journal of Nutritional Science and Vitaminology, Tokyo, 28*, 491–500.

Oakley, G. P., Erckson, J. D., James, L. M., Mulinare, J., & Cordero, J. F. (1994). Prevention of folic acid preventable spina bifida and anencephaly. In G. Bock & J. Marsh (Eds.), *Neural tube defects* (pp. 212–222). Cheicester: Wiley.

Odutaga, A. A. (1982). Effects of low-zinc status and essential fatty acid deficiency on growth and lipid composition of rat brain. *Clinical Experimental Journal of Pharmacology and Physiology, 9*, 213–221.

Otte, C., Zhao, S., & Whooley, M. A. (2012). Statin use and risk of depression in patients with coronary heart disease: Longitudinal data from the Heart and Soul Study. *Journal of Clinical Psychiatry, 73*, 610–615.

Perry, M. L., Gamallo, J. L., & Bernard, E. A. (1986). Effect of protein malnutrition on glycoprotein synthesis in rat cerebral cortex slices during the period of brain growth spurt. *Journal of Nutrition, 116*, 2486–2489.

Perry, L. A., Stigger, C. B., Aimnsworth, B. E., & Zhang, J. (2009). No association between cognitive achievements, academic performance, and serum cholesterol concentrations among school-aged children. *Nutritional Neuroscience, 12*, 160–166.

Prince, M., Prina, M., & Guerchet, M. (2013). *Journey of caring: An analysis of long-term care for dementia* (World Alzheimer Report 2013). Available at: http://www.alz.co.uk/worldreport 2013. Accessed 24 Mar 2015.

Qawasmi, A., Landeros-Weisenberger, A., Leckman, J. F., & Bloch, M. H. (2012). Meta-analysis of long-chain polyunsaturated acid supplementation of formula and infant cognition. *Pediatrics, 129*, 1141–1149.

Reddy, T. S., & Ramakrishnan, C. V. (1982). Effects of maternal thiamine deficiency on the lipid composition of rat whole brain, gray matter and white matter. *Neurochemistry International, 4*, 495–499.

Reisbick, S., Neuringer, M., Gohl, E., Waid, R., & Anderson, G. J. (1997). Visual attention in infant monkeys: Effects of dietary fatty acids and age. *Developmental Psychology, 33*, 387–395.

Repo-Tiihonen, E., Halonen, P., Tiihonen, J., & Virkkunen, M. (2002). Total serum cholesterol level, violent criminal offences, suicidal behavior, mortality and the appearance of conduct disorder in Finnish male criminal offenders with antisocial personality disorder. *European Archives of Psychiatry and Clinical Neuroscience, 252*, 8–11.

Rosenzweig, M. R., & Leiman, A. L. (1989). *Physiological psychology*. New York: Random House.

Schreurs, B. G. (2010). The effect of cholesterol on learning and memory. *Neuroscience and Behavioral Reviews, 34*, 1366–1379.

Shukla, S. D., & Halenda, S. P. (1991). Phospholipase D in cell signaling and its relationship to phospholipase C. *Life Sciences, 48*, 851–866.

Solomon, A., Kivipelto, M., Wolozin, B., Zhou, J., & Whitmer, R. A. (2009). Midlife serum cholesterol and increased risk of Alzheimer's and vascular dementia three decades later. *Dementia and Geriatric Cognitive Disorders, 28*, 75–80.

Spector, A. A., & Yorek, M. A. (1985). Membrane lipid composition and cellular function. *Journal of Lipid Research, 26*, 1015–1035.

Tacconi, M. T., Calzi, F., & Salmona, M. (1997). Brain lipids and diet. In M. Hillbrand & R. T. Spitz (Eds.), *Lipids, health, and Behavior* (pp. 197–226). Washington, DC: American Psychological Association.

Wallner, B., & Machatschke, I. H. (2009). The evolution of violence in men: The function of central cholesterol and serotonin. *Progress in Neuro-Psychopharmacology & Biological Psychiatry, 33*, 391–397.

West, R., Beeri, M. S., Schmeidler, J., Hannigan, C. M., & Angelo, G. (2008). Better memory functioning associated with higher total and low-density lipoprotein cholesterol levels in very elderly subjects without the apolipoprotein e4 allele. *The American Journal of Geriatric Psychiatry, 16*, 781–785.

Whalley, L. J., Deary, I. J., Starr, J. M., Wahle, K. W., & Rance, K. A. (2008). n-3 Fatty acid erhthrocyte membrane content, APOE varesilon4, and cognitive variation: An observational follow-up study in late adulthood. *American Journal of Clinical Nutrition, 867*, 449–454.

Yurko-Mauro, K., McCarthy, D., Rom, D., Nelson, E. B., & Ryan, A. S. (2010). Beneficial effects of docosahexaenoic acid on cognition in age-related cognitive decline. *Alzheimer's & Dementia, 6*, 456–464.

Zhang, J. (2011). Epidemiological link between low cholesterol and suicidality: A puzzle never finished. *National Neuroscience, 14*, 268–287.

Chapter 5
Neurotransmitters and Short-Term Effects of Nutrition on Behavior

Abstract The chapter, at length, would discuss neurotransmitters and precursor control, the role of serotonin, sleep pattern, mood fluctuations, the cycle of food intake, serotonin and nutrient selection, diet, serotonin and mood fluctuations, carbohydrate craving and mood fluctuations and role of acetylcholine in human health. Additionally, the chapter would also discuss physiologic diseases like Huntington's Chorea, Tardive Dyskinesia and Alzheimer's Disease. Additionally, the chapter would also discuss Dopamine and Norepinephrine, Tyrosine and stress and also Effect of Tyrosine on Mood fluctuations.

Keywords Neurotransmitter · Precursor control · Serotonin · Sleep pattern · Mood fluctuations · Huntington's chorea · Tardive dyskinesia · Alzheimer's disease · Tyrosine · Mood

5.1 Neurotransmitters and Precursor Control

In the central nervous system (CNS), there are close to 60 substances which are believed to act as neurotransmitters. The three major types according to Schwartz (2001) are:

Biogenic amines: These include substances such as serotonin, histamine, catecholamines (dopamine and epinephrine), and the ester acetylcholine.
Neuropeptides: These include substances such endorphins, cholecystokinin, neurokinin A, and somatostatin.
Amino acids: These include glutamate, glycine, and GABA (γ-aminobutyric acid).

Neurotransmitters may also be classified on the basis of their function. Excitatory neurotransmitters such as epinephrine increase the likelihood of an action potential in the postsynaptic cell. By contrast, inhibitory neurotransmitters, such as serotonin, serve to calm rather than stimulate the brain. Of the three major types, the synthesis and activity of biogenic amines is the most thoroughly researched, because they have a dietary importance, which have a pronounced behavioral impact. They are of low molecular weight, are water soluble, and carry an ionic charge. They are synthesized in the neurons from precursor molecules which ordinarily must be obtained in whole or in part from diet. Precursor control means that the availability of the neurotransmitter is controlled by the presence of its precursor or the substance from which it is synthesized. For example, tryptophan is the precursor to the neurotransmitter serotonin. The cells of the body cannot produce tryptophan; hence, individuals must consume it in sufficient amounts through the diet. In contrast, choline is the precursor of the neurotransmitter acetylcholine, but it can be formed in the liver and brain. However, the major portion of choline is obtained through dietary lecithin. Under appropriate conditions, increasing the dietary intake of a precursor should stimulate formation of neurotransmitter. However, a number of conditions must be met before it can be assumed that the rate at which neurons will synthesize a given neurotransmitter is dependent on the intake of a dietary precursor (Wurtman et al. 1980). They are the following:

It must be demonstrated that the precursor is obtained from the general circulation and cannot be synthesized in the brain.

Plasma levels of the precursor must fluctuate with dietary intake and not be kept within a narrow range by a physiological mechanism.

The enzyme transforming the precursor into the neurotransmitter must be unsaturated so that the synthesis of the neurotransmitter will accelerate when additional precursor material is made available.

The enzyme that catalyzes the synthesis of the neurotransmitter is not modified by feedback inhibition that could decrease its activity after the neurotransmitter levels have increased.

The rate at which the precursor enters the brain must vary directly with its concentration in plasma – there must be an absolute blood-brain barrier for the precursor. As its name suggests, the blood-brain barrier is the structural means by which most substances in the blood are prevented from entering the brain. The barrier is semipermeable, allowing some materials to cross into the brain but stopping others from doing so. Many special characteristics contribute to this ability. To begin with, a discontinuous sheath of astrocyte cells which are interspersed between blood vessels and neurons envelop the cerebral capillaries or small blood vessels of the brain. As in most other parts of the body, the capillaries are lined with endothelial cells. In muscles, for example, the endothelial tissue has small spaces in between each individual cell so that substances can move easily and readily between the inside and outside of the capillaries. However, the endothelial cells of the capillaries in the brain differ ultrastructurally from those in the muscle. That is, they fit together with extremely tight junctions, which serves to prevent substances from passing out of the bloodstream. Finally, the plasma membranes of the cerebral endothelial cells provide a continuous lipid barrier between blood and brain and are at the anatomical basis of the barrier (Abbot et al. 2006).

The most important function of the blood-brain barrier is to help maintain a constant environment for the brain. It serves to protect the brain from foreign substances in the blood which might injure the brain. Also, it protects the brain from the rest of the hormones in the body. Large molecules do not easily pass through the barrier, and those that have a high electrical charge are slowed down. However, lipid-soluble molecules, such as alcohol or barbiturate drugs, can pass rapidly into the brain. The physiological consequences of this are, invariably, adverse.

5.2 Serotonin

Serotonin is also called 5-hydroxytryptamine or 5-HT is accomplished within the serotonergic neurons located in the brain. Tryptophan is an amino acid, one of the building blocks of protein. The synthesis of serotonin is graphically illustrated in Fig. 5.1 (below).

Hydroxylation of tryptophan to 5-hydroxytryptophan is the primary factor limiting the synthesis of serotonin, the reaction being catalyzed by tryptophan hydroxylase. Investigations of tryptophan hydroxylase on animals have shown that this enzyme is only half saturated at the concentrations of tryptophan normally found in the rat brain (Sved 1983). This would suggest that increasing the availability of tryptophan could double the rate of tryptophan synthesis. In their path-breaking research, in the early 1970s, Fernstrom and Wurtman attempted to do just this. Rats given varying amounts of tryptophan showed a clear dose response. As brain tryptophan levels increased, serotonin demonstrated a corresponding rise. Serotonin levels are therefore sensitive to tryptophan levels, with even small changes in brain levels of tryptophan producing significant effect on 5-HT (Young 2005).

Fig. 5.1 Synthesis of serotonin

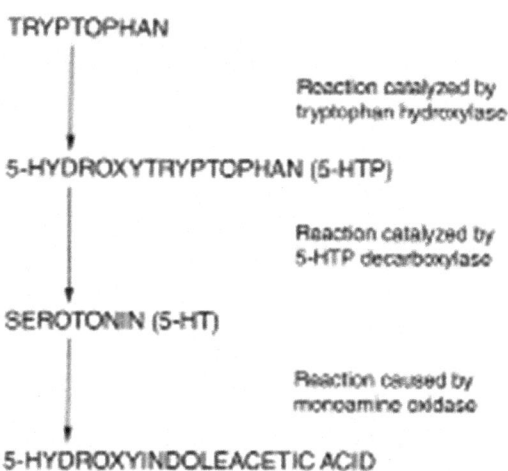

Enhancing plasma tryptophan elevates brain tryptophan levels and accelerates the synthesis of serotonin. The earlier cited authors predicted similar results following the consumption of a high-protein meal that naturally contained tryptophan by rats. But, much to their surprise, brain tryptophan and serotonin levels were depressed, although plasma tryptophan level enhanced. This paradox was found to occur because tryptophan is relatively scarce in protein when compared with other large neutral amino acids (LNAAs). The LNAAs, namely, tryptophan, valine, tyrosine, leucine, isoleucine, phenylalanine, and methionine, share a common transport mechanism in crossing the blood-brain barrier (Pardridge 1986). Since the LNAAs share this common transport system, all of them compete with each other to enter the brain. The amount of a given LNAA that is transported into the brain therefore depends on the level of that amino acid in the blood relative to the other LNAAs. And, since tryptophan comprises only about 1–1.5% of any dietary proteins, it is not a very good competitor (Sved 1983). Following a high-protein meal, the plasma levels of other LNAAs increase to a greater degree than plasma tryptophan levels. Thus, a high-protein meal will give the other LNAAs a competitive advantage in crossing the blood-brain barrier.

The discovery that in contrast to a high-protein meal, it is a high-carbohydrate meal that elevated the brain levels of tryptophan and serotonin is a surprising paradox. As the high-carbohydrate test meals contained no protein or tryptophan, this discovery demanded a scientific explanation. It was soon discovered that brain levels of tryptophan and serotonin rose in fasted animals because carbohydrate intake stimulated secretion of insulin. In fact the investigation of Fernstrom and Wurtman revealed that in rats tested with high-carbohydrate diets as against insulin injections, similar results were obtained in both experimental groups – enhancement of plasma and brain tryptophan and serotonin level. As it turns out, plasma tryptophan has the unusual ability to bind itself loosely to circulating albumin. When insulin is administered or secreted, nonesterified fatty acid molecules, which are usually bound to albumin, will dissociate themselves and enter adipocytes. As a result of this, tryptophan that binds to the albumin is protected from being taken up by peripheral cells. The result of this action is that little change occurs in plasma tryptophan levels following insulin secretion, but the plasma levels of many other LNAAs decrease because the bound tryptophan is nearly able to cross into the brain as is the unbound tryptophan which is still in circulation; insulin spares tryptophan levels rather than interfering with its transport (Yuwiler et al. 1977). Further, insulin decreases plasma levels of other LNAAs by stimulating their uptake into the muscle. The increase in the total tryptophan:LNAAs ratio that results causes an increase in the amount of tryptophan which crosses the blood-brain barrier. It is important to note that only a small amount of high-quality protein, as little as 4%, is sufficient to block the effects of carbohydrate on brain tryptophan levels. A diet consisting of both carbohydrate and protein will increase the plasma levels of tryptophan, but the brain levels of tryptophan and subsequent uptake of 5-HT will decrease (Yokogoshi and Wurtman 1986). Additionally, the extent to which a high-carbohydrate meal can raise tryptophan levels depends on whether other foods are present in the individual's stomach. For instance, if sufficient protein remains in one's stomach from the last meal, the effects of carbohydrate on brain tryptophan and serotonin will be blunted. Hence, it is unlikely that a typical meal consumed under normal circumstances, even if high in carbohydrate, will significantly boost serotonin levels (Young 2005). Further, the time the meal is consumed will also affect serotonin levels.

5.3 Diet, Serotonin, and Behavior: The Triad of Food and Behavior Link

Serotonin is believed to play a key role in the mediation of mood and antisocial behavior, such as aggression. It also affects sleep and arousal, as well as food intake and nutrient selection. Given the impact of carbohydrate intake on serotonergic system, diet-induced alterations in serotonin could well serve to modulate behavior. There is evidence that supports and contradicts this proposition.

5.4 The Cycle of Food Intake, Serotonin, and Sleep Pattern

Investigations with animals have shown that serotonin plays an important part in sleep pattern. While higher serotonin levels induce sleepiness, lower levels cause insomnia. For instance, giving enough tryptophan to increase brain 5-HT levels by 10–20% decreases the latency to sleep in rats. In cats, sleep was significantly reduced when the neurons of the raphe nuclei were destroyed (Jouvet 1968). The raphe nuclei are an area of the hypothalamus, rich in 5-HT, and their destruction depressed cerebral levels of serotonin. Tryptophan can also increase sleep in normal human beings and has been recommended for the treatment of insomnia. Small doses have been shown to shorten sleep latency, increase sleep duration, and decrease the frequency of waking up during sleep in both men and women.

As carbohydrate-enriched food would enhance brain serotonin, it should follow that carbohydrate intake should increase feelings of sleepiness. This would imply that individuals who eat carbohydrate-rich food feel less alert unlike those who eat protein-rich food (Spring et al. 1989). Males characterized themselves as calmer after consuming carbohydrate-rich food than after protein-rich food, while females described themselves as sleepier, irrespective of the fact whether it was breakfast or lunch, unaffected by age of the individuals. These results suggest that a high-carbohydrate-rich meal may, in fact, increase the level of brain serotonin, given the reduction in mental alertness. From an anthropological perspective, the siesta after a heavy carbohydrate (rice)-laden meal in hot countries such as India goes to show that perhaps it is the enhanced serotonin content in the brain that triggers slowed mental alertness followed by sleep, which has become a custom in these regions.

As an alternative to medication, tryptophan has been prescribed for the treatment of insomnia (Hudson et al. 2005). In 1989 a mysterious malady struck thousands of people in the USA due to tryptophan supplements. Exhibiting flu-like symptoms, this neurological condition was called eosinophilia-myalgia syndrome (EMS), leading to 37 deaths and another 15,000 permanently disabled. This led to a ban in the sale of tryptophan in the USA in 1991. Epidemiological detective work subsequently traced the outbreak to a sole Japanese manufacturer whose tryptophan supplements may have contained impurities tied to genetically engineered (GM) bacteria (Mayeno and Gleich 1994), although it has been more recently hypothesized that the metabolites produced by large dosages of tryptophan inhibit the degradation of histamine, which in excess could cause EMS (Smith and Garrett 2005). Since 2002 the US FDA has loosened the tryptophan sale by prescription and also made it available to the public through mail order retailers.

The effect of tryptophan in sleep induction cannot be generalized, as there are other investigations which contradict the observation. It should be recognized that observation on sleep induction immediately after ingestion of tryptophan need not necessarily show a positive effect; perhaps waiting for an hour to record an observation might be necessary.

5.5   The Cycle of Food Intake, Serotonin, and Nutrient Selection

In contrast to investigations on serotonin and sleep, there is a large body of evidence surrounding the neurons which produce serotonin which are important in regulating food intake. Reduced food intake and enhanced activity of serotonergic neurons has been observed. The inverse relationship between food intake and reduced serotonergic activity has also been reported. Serotonergic neurons appear to play a role in overall energy consumption, as well as macronutrient intake. The evidence for this comes from investigations which examined the changes in nutrient choice following experimental alterations in serotonergic systems. Most of these experiments have assessed the effect of chemicals like tryptophan, fenfluramine, or fluoxetine on nutrient intake in animals, for example, rats are given a choice of diets which contain varying proportions of the macronutrients, protein, carbohydrate, and fat. Initial experiments that used this paradigm found that increases in serotonergic activity were accompanied by selective reduction in carbohydrate intake. Subsequent research, however, has indicated that a number of independent variables, such as the age of animals, diet composition, and feeding schedule, must be considered when assessing the effects of serotonin on nutrient choice. When these variables are taken into account, an elevation in serotonin level may lead to certain reductions in carbohydrate intake, though it is not a very robust phenomenon (Kanarek 1987).

Although it has been assumed that CNS mediates the effects of serotonin on food intake and diet selection, adjustments in food intake may also result from changes in peripheral serotonergic systems. In fact, the largest stores of serotonin are not in the brain, but in the gastrointestinal tract. Hence, modifying the peripheral 5-HT, without changing its levels in the brain, can yet affect feeding behavior. For example, peripheral administration of serotonin that does not cross the blood-brain barrier, nevertheless, inhibits total energy intake and selectively reduces fat intake (Amer et al. 2004). These results indicate that both the CNS and peripheral serotonergic system must be considered when evaluating the relationship of serotonin and feeding behavior. In humans, manipulations in serotonergic activity, which can be brought about by drugs like fenfluramine which increases serotonergic activity, will bring about reduction in energy intake in obese people, possibly through reduction in meal size and slowing food intake (Rolls et al. 1998).

A clear way to increase cerebral serotonin levels is to administer tryptophan levels among willing subjects. Such investigations using this method have found that large doses of tryptophan reduce energy intake in individuals of normal weight. In general, adding tryptophan to the diet leads to equivalent reductions in carbohydrate and protein intake. However, there does exist data which indicates that tryptophan combined with a high-protein meal can selectively decrease the amount of carbohydrate subsequently consumed (Blundell and Hill 1987). In a study that used the reverse approach, i.e., acute tryptophan depletion, Young and his colleagues provided their male volunteers with meals which were deficient in tryptophan. Following an overnight fast, the subjects received either a nutritionally balanced mixture for breakfast or one which was tryptophan deficient. Five hours later, the men were provided an opportunity to eat all that they wished to eat, for lunch, from an experimenter-furnished buffet. The results indicated that relative to the men who ate a balanced breakfast, those who ate the tryptophan-deficient breakfast ingested significantly less protein at lunch (Young 1991). Carbohydrate intake did not appear to be affected. In contrast, using the acute tryptophan depletion method with a sample of overweight adults, it was shown that lowering tryptophan via breakfast beverage resulted in the subjects increasing their intake of sweet-tasting foods. Taken together, these studies suggest that increases in serotonergic activity are associated with reductions in energy intake in humans and animals. While it is premature to determine if increases in 5-HT lead to selective reduction in carbohydrate intake, some research does suggest that the brain may signal less of a need for certain macronutrients if tryptophan levels are manipulated.

5.6   Diet, Serotonin, and Mood Fluctuations

There is now conclusive evidence that serotonin definitely affects mood disorders (Lopez –Figueroa et al. 2004). That quality of life is affected by mood disorder is a well-established fact. Of all this, depression is the most challenging, requiring psychiatric counseling. In truth, the etiology of depressive illness cannot be explained by a single biological, social, or developmental theory. A number of factors must interact before a mental disorder precipitates. At least in some patients, alterations in brain serotonin lead to mood fluctuations. As pharmacological research has shown, drugs that enhance serotonergic activity are useful in treating depression and drugs that decrease serotonergic activity will precipitate depression. Though it might be an oversimplification to deduct that one single serotonergic factor is at the root of depression, it is safe to conclude that brain serotonin is somehow related to mood swings. Results with food intake and mood swings are rather mixed in healthy individuals. For instance, tryptophan-depleted beverages are found to enhance depression in healthy males (Young et al. 1985) and females (Ellenbogen et al. 1996). While it has been shown that a carbohydrate-rich meal can induce calmness in healthy adults (Spring et al. 1989), others have not reported any such effects. Studies have also been conducted among depressed individuals to know the effect of diet on serotonin as it may relate to their depression. In the first type, researchers examine the plasma tryptophan levels of depressed and nondepressed individuals and look for differences in tryptophan levels by group, predicting lower levels in depressed individuals. These investigations have established that depressed individuals have a lower tryptophan:LNAA ratio than nondepressed individuals (Lucca et al. 1992). This in effect, as has been seen earlier, is due to the inability of tryptophan to cross the blood-brain barrier. In the second type, researchers use the acute tryptophan depletion method with depressed individuals to determine the effect on their subsequent mood. For these trials, the suitable subjects are patients who have gone into remission after treatment with antidepressant drugs. For example, patients are given a tryptophan-depleted beverage versus that which includes a mixture of other LNAAs, and the maintenance of remission is observed. Some researchers have found that twice as many subjects who receive the tryptophan-depleted beverage relapse, relative to the LNAA subjects (Delgado et al. 1991), but others indicate that this is likelier the case for individuals who have family history of depression (Ruhe et al. 2007). Except when severe, tryptophan may be a useful antidepressant, taking into consideration all available evidence (Young 2005). Evidence on the effect of tryptophan as an enhancer or reliever of aggression is rather ambivalent.

5.7   Carbohydrate Craving and Mood Fluctuations

The possible relationship between carbohydrate intake and behavioral changes has been the focus of research during the past quarter century. In terms of depression, one of the forms has been linked to appetite, the condition known as seasonal affective disorder (SAD). SAD symptoms appear to increase from late fall or early winter and decline in the spring as the days get noticeably longer. This phenomenon may not apply to countries on the Asian, African, and Latin American continents, as no such properly demarcated seasons, such as fall, winter, spring, and summer, apply to these continents, as in the case of the West, like the USA, Europe, and Scandinavia. And, situation in countries like South Africa and Australia varies very

differently, as these countries experience the polar climate, for example, when summer sets in the USA, Europe, Asia, Africa, or Latin America, winter sets in these
two countries, and when winter sets in Australia and South Africa, the reverse hap-
pens in other countries mentioned above. Besides the characteristics of seasonality,
those who suffer from SAD symptoms experience increased cravings for carbohydrate and enhanced sleep during this period. Some of these individuals anecdotally
report that they consume carbohydrates in order to combat depression (Leibenluft
et al. 1993). An association between depressive symptoms and carbohydrate craving
has also been noted among women with premenstrual symptoms (PMS). Acute
tryptophan depletion has also been shown to aggravate PMS, in particular irritability (Menkes et al. 1994). Additionally, there appears to be a subgroup of obese
women who describe frequent and powerful cravings for carbohydrate-rich foods
(Wurtman and Wurtman 1995). Many of these exhibit negative mood states, such as
being tense or restless, prior to consumption of a snack, but relaxed after snacking.
For these obese women, high-carbohydrate/high-fat snacks are preferred, and the
addition of upward of 800 calories per day may very well contribute to the maintenance of their excessive weight.

It has been suggested that the strong desire for carbohydrates exhibited by individuals with SAD and PMS and carbohydrate craving among the obese may reflect
the body's need to increase serotonin levels, which, in turn, will lead to a positive
mood change when the need is met. This could be a body mechanism of these individuals where carbohydrate craving is preceded by depression, which is alleviated
subsequently, in fact a self-medication process (Christensen 1996).

While it is intriguing why individuals consume specific food to modulate their
mood fluctuations, it also presents numerous problems. For instance, an objective
definition of what constitutes an excessive consumption of carbohydrate has not
been established. Most of such investigations have been done with individuals who
are known as carbohydrate cravers. Thus, the designation of individuals as carbohydrate cravers has been based on their own perceptions of their behavior, rather than
on objective criteria. Further, the definition of a high-carbohydrate food has been
rather problematic. It is also true that high-carbohydrate foods are also rich in fat,
such as cake, chocolates, and ice cream. Given the high palatability of such food,
cravings for such food may indicate a desire for a pleasant gustatory experience,
rather than for a specific nutrient. Addressing these shortcomings, a recent investigation carefully defined carbohydrate craving, as well as supplying a carbohydrate-
rich food and measuring actual carbohydrate intake. After inducing a dysphoric
mood state, with sad music and recalling negative thoughts, the investigators offered
carbohydrate cravers either a carbohydrate-rich or a protein-rich beverage in a
double-blind placebo controlled manner. On the third day, the participants chose
between the two beverages themselves. Based on their reactions, over the previous
2 days, the participants chose the carbohydrate-rich beverage significantly more
often and reported that it produced a greater improvement in their mood (Corsica
and Spring 2008).

It also transpires that some physiological data indicate that dietary-induced alterations in the plasma ratio of tryptophan to the other LNAAs may have less effect on

brain serotonin and function than was previously thought. For example, receptors for serotonin are located in the dendrites and cell body of the serotonergic neurons. Feedback from these receptors serves to limit the dietary-induced effects of alterations in neurotransmitter activity (Nicholas and Nicholas 2008). The injection of tryptophan thus leads to a reduction in the firing of the serotonergic neurons, which serves to counteract the increased release of serotonin which might have otherwise occurred. In addition, recall that even minimal amounts of protein in a high-carbohydrate food can suppress that food's ability to raise brain tryptophan and serotonin levels. Taken together, these problems still serve to challenge the full-scale acceptance of the hypothesis that individuals who suffer from SAD and PMS or carbohydrate craving obesity seek to consume high-carbohydrate foods in order to modulate their mood (Fernstrom 2000).

5.8   The Role of Acetylcholine in Human Health

Acetylcholine is another important neurotransmitter which is synthesized from acetyl coenzyme-A and choline. Like tryptophan, choline enters the brain via a transport system which allows it to cross the blood-brain barrier. Once it enters the brain, choline is taken up by the neuron, where in the presence of the enzyme choline acetyltransferase the acetate ion is transferred from acetyl coenzyme-A (CoA) to the choline molecule as depicted in the following figure (Fig. 5.2).

> ACETYL CoA + CHOLINE – Catalysis by choline acetyltransferase
> – ACETYLCHOLINE + CoA

The synthesis of acetylcholine is influenced by the availability of choline within the cholinergic neuron. Choline can be synthesized in the liver, but neuronal choline concentration can also be altered by dietary choline intake, in the form of either free choline or as a constituent of the phospholipid lecithin (Hirsch and Wurtman 1978). Lecithin, or phosphatidylcholine, is present in a variety of foods, including eggs, poultry, fish, liver, peanuts, and wheat germ, which is also an ingredient of many processed foods where it serves as an antioxidant and emulsifying agent. In addition, lecithin is also available as a dietary supplement and is sold as such in pharmacies supplying vitamins. The intestinal mucosa easily absorbs the lecithin from food or these supplements which is rapidly hydrolyzed by free choline. Any lecithin

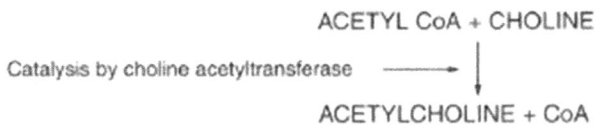

Fig. 5.2 Synthesis of acetylcholine

which is not hydrolyzed enters the bloodstream and is transported to the lymphatic circulation system where it is broken down more slowly to choline.

The plasma levels of choline will be enhanced by the intake of foods that contain substantial amounts of lecithin or choline. Since choline can easily cross the brain-blood barrier, elevations in choline levels translate into enhanced brain levels of choline. Further, since the enzyme choline acetyltransferase is unsaturated when choline is within normal limits, increased neuronal levels of the precursor will stimulate the synthesis of acetylcholine. Diet-induced increases in neuronal acetylcholine levels are therefore associated with enhanced release of choline when cholinergic neurons are stimulated (Wecker 1990).

Many neurological diseases, such as Huntington's chorea, tardive dyskinesia, and Alzheimer's disease, are thought to involve deficiencies in the activity of cholinergic neurons (Fernstrom 2000). Drugs which serve to enhance cholinergic transmission are typically used in the treatment of these diseases. However, their effectiveness has been inconsistent. Additionally, most of these drugs have only a short-duration effect and produce unpleasant side effects such as vomiting, nausea, and mental dullness, which further limit their application. It has therefore been proposed that increasing acetylcholine through dietary manipulations may be a more effective and benign strategy for the treatment of diseases associated with deficiencies in cholinergic neurotransmission. The following details dwell on the utility of acetylcholine therapy in containing some of the listed diseases:

5.9 Huntington's Chorea, Tardive Dyskinesia, and Alzheimer's Disease

5.9.1 Huntington's Chorea

Huntington's chorea is an inherited progressive neurological disorder whose symptoms usually begin when an individual reaches middle age. The characteristic symptoms include involuntary muscular contractions known as chorea which involve all the skeletal muscles. Chronic chorea results in poor balance, difficulty in walking, and restlessness. As the disease progresses, signs of mental disturbance develop, such as confusion, forgetfulness, inability to concentrate, personality changes, paranoia, and dementia. Autopsies performed on deceased patients who suffered from Huntington's disease reveal that their brain weights were reduced, particularly in regions of the cortex and basal ganglia.

By examining neurotransmitter functioning in patients with the disease, some evidence has been provided for a role for acetylcholine. For instance, patients with the disease show reduced levels of acetylcholine as well as choline acetyltransferase and decreased numbers of postsynaptic receptors. When drugs are used to increase acetylcholine activity, the choreic movements characteristic of the disorder appear to be reduced, while drugs that increase activity of the neurotransmitter tend to aggravate them.

It has thus been suggested that administration of choline or lecithin would increase levels of acetylcholine and thereby reduce some of the disease symptoms. In fact, a few investigators have reported that treatment with choline significantly improved balance, gait, and choreic movement in some patients (Aquilonius and Eckeras 1977). However, their use has generally been unsuccessful, as improvements when found did not persist for more than 1 or 2 weeks, even when choline was administered for a continued period (Rosenberg and Davis 1982).

5.9.2 Tardive Dyskinesia

This disease is the unfortunate side effect of certain antipsychotic medications. It is a neurological disorder which can develop after prolonged treatment of 3 months or more of drugs such as haloperidol or chlorpromazine. Individuals who have experienced sustained exposure to antipsychotic medications, electroconvulsive therapy, or organic brain syndrome, as well as histories of prolonged drug or alcohol abuse, are at particular risk (Kompoliti and Horn 2007). The disease is not an inevitable consequence of using antipsychotic medication and even less so with the second generation of antipsychotic medications, but may develop in 5–15% of patients who use the drugs for more than 1 year (Corell et al. 2004).

The disease is characterized by hyperkinetic activity of the mouth and jaw region, protrusion of the tongue, lip smacking and puckering, and difficulty in swallowing. In addition, involuntary spastic movements of the hands, arms, feet, and legs may be present. Why the disease develops in certain individuals is unknown, though it is suspected that an imbalance between the cholinergic and dopaminergic neurons in the basal ganglia may be involved. This imbalance would seem to favor the transmission of the neurotransmitter dopamine at the expense of the acetylcholine being transmitted. Research with drugs shows that enhancing the activity of acetylcholine relieves the symptoms of the disease, while drugs which reduce its activity will exacerbate the individual's symptoms. Thus, some support exists for the role of decreased cholinergic activity in tardive dyskinesia.

In a number of investigations, the use of choline or lecithin has been associated with reducing the frequency of abnormal movements of some patients with the disease. In a double-blind crossover study, it was found that choline administration increased plasma choline levels and suppressed involuntary facial movements in nearly half of their patients with the disease (Growdon et al. 1977). A problem with their study, however, was the telltale odor that characterized the patients who ingested the choline. The treatment group developed the aroma of rotten fish in their urine and perspiration, as well on their breath, which made the likelihood of the study being truly blind in character, somewhat questionable. The odor was produced by the action of intestinal bacteria and choline and does not occur after lecithin administration. Subsequent investigations which used lecithin instead of choline indicate that lecithin is as effective as choline in suppressing the disease (Jackson et al. 1979). However, many patients reported feeling of nausea or experience abdominal cramps and diarrhea while taking high doses of lecithin. Because of these side effects, and more recent work that has shown choline to be less effective than first believed, current recommendation is that the dietary treatment of the disease is of only limited clinical utility (Gelenberg et al. 1989).

5.9.3 Alzheimer's Disease

Recent years have shown a widespread occurrence of the Alzheimer's disease, which affects one in eight adults over the age of 65 and nearly half of those over 85 years of age. The disease is characterized by the slow, progressive deterioration of cognitive functions, with dementia usually occurring within 5–10 years. The onset of the disease is usually inconspicuous and very often difficult to distinguish from other psychological problems. Initially, the patient suffers from periods of short-term memory loss and may find it hard to concentrate. As the disease progresses, however, memory loss becomes more severe, while disorientation, anxiety, depression, and difficulty in completing simple tasks become more frequent.

The deficits associated with the disease fall into three domains, which encompass decline in cognition, everyday functioning (the ability to perform activities of daily living), and behavior. Cognitive losses in individuals with the disease include forgetfulness, memory loss, lack of concentration, and declines in language ability. Since the disease affects executive functions, the ability to perform basic life tasks can be compromised. Thus, in the functioning domain, motor skills are compromised, which can affect the ability to walk and talk. Finally, behavioral changes, such as mood swings, depression, and irritability, also accompany the disease. Agitation and the incidence of aggressive behavior (screaming, hitting, biting, etc.) increase with the disease progression and place a particularly heavy responsibility on caregivers.

The cognitive declines that accompany the disease occur as neurons in the brain die and communication between the cells breaks down, but may also be due, in part, to alterations in cholinergic activity in the CNS. Acetylcholine neurons in the hippocampus play an important role in memory, and postmortem examination of the patients affected by the disease reveals significant reduction in choline acetyltransferase levels in the hippocampus and neocortex. Moreover, dementia and low choline acetyltransferase levels appear to be positively correlated. Research suggest that there may be selective degeneration of cholinergic nicotinic receptors in certain regions of the brain, such as the nucleus basalis (Ereshefsky et al. 1989).

Researchers who study memory have determined that anticholinergic agents will disrupt performance on tests that rely on the recall of words and digits (Chew et al. 2008). Since problems with memory are a major symptom of the disease, much attention has been paid to the question of whether the consumption of choline or lecithin can bring about improvements in memory or alleviate deficits. With animal models, the administration of choline has been shown to elevate brain acetylcholine levels and to reduce age-related declines in retention of learned behavior. In humans, lecithin consumption does raise choline levels in plasma (Wurtman et al. 1977), and, in at least one study with normal young adults, choline produced improvements in memory performance (Sitaram et al. 1978).

However, in samples from elderly persons for which the memory problems that accompany the disease may represent a significant clinical problem, acetylcholine precursors have produced disappointing results, as neither choline nor lecithin seems to offer much improvement in memory (Spring 1986). Work with normal, healthy subjects has not shown an effect of acetylcholine on improving memory either (Nathan et al. 2001). Furthermore, it has also been noted that depressed mood is a side effect of choline or lecithin treatments employed in reverse memory deficits in Alzheimer's patients (Davidson et al. 1991).

Despite the mixed results with Alzheimer's patients, some researchers are still convinced that the exploration of cholinergic mechanisms in affective disorders, such as mania, may offer important clues concerning the physiology behind psychopathology (Janowsky and Overstreet 1998). For instance, the brains of patients with major depression do show higher levels of choline than the brains of control subjects (Charles et al. 1994). Advances in neuroimaging and molecular genetics may one day help in identifying cholinergic linkages to mood. In sum, it must be concluded that, based on investigations, to date, the administration of choline or lecithin in older adults appears to have only limited benefits in the treatment of neurological disorders, such as Huntington's disease, tardive dyskinesia, or Alzheimer's disease.

5.10   Dopamine and Norepinephrine

In the mammalian brain, including humans, the catecholamine neurotransmitters dopamine and norepinephrine are synthesized from the amino acid tyrosine. The synthesis of dopamine occurs within the dopaminergic neurons, while the synthesis of norepinephrine occurs within the noradrenergic neurons. As shown below (Fig. 5.3), the rate-limiting initial step in this synthetic pathway is the conversion of tyrosine into dihydroxyphenylalanine by the enzyme tyrosine hydroxylase. This intermediate product, called DOPA, is then synthesized into dopamine by the action of the enzyme DOPA decarboxylase. In noradrenergic neurons that also contain the enzyme dopamine-β-hydroxylase, dopamine is then converted into norepinephrine.

Tyrosine like the neurotransmitter tryptophan is obtained from diet, but facilitated by protein sources. As serotonin increases when the relative level of plasma tryptophan increases, so does catecholamine synthesis increases as the relative level of tyrosine increases. Tyrosine meets most of the precursor conditions that were listed earlier, as its plasma levels increase following protein intake, its rate-limiting enzyme is unsaturated, and a transport system exists to ferry it across the blood-brain barrier (Christensen 1996). While nearly all conditions necessary for tyrosine to increase the synthesis of its neurotransmitter products are met, most studies have

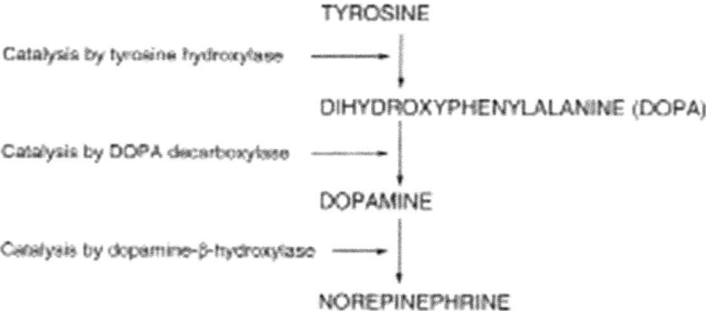

Fig. 5.3 Synthesis of DOPA and norepinephrine

revealed that tyrosine administration has no effect on the synthesis of catecholamines. Indeed, not only have most studies indicated that tyrosine administration is ineffective to increase DOPA, but brain levels of the metabolites of the catecholamines have not been shown to increase (Sved 1983).

5.11   Tyrosine and Stress

When laboratory animals are exposed to acute, unavoidable (to be honest, very cruel in an ethical sense) stress, it will lead to their behavioral deficits, including reductions in spontaneous motor activity and aggression. These behavioral changes are described as learned helplessness and have been attributed, by some researchers, to hippocampal and brainstem depletions in norepinephrine (Petty et al. 1993). It is important to know whether tyrosine would reverse both the behavioral deficits and the norepinephrine depletion which follows an uncontrolled stressor. Some research with mice suggests that this does happen, as a cold swim stress test decreased brain norepinephrine and dopamine levels, but tyrosine prevented the reduction in aggression and activity that ordinarily would have been observed.

Research has shown that in the absence of stress, administering tyrosine does nothing to affect performance levels in humans, a finding which supports the animal research (Owasoyo et al. 1992). But, when male subjects were subjected to environmental stressors, like cold or high altitude, tyrosine supplements served to counteract the detrimental effects and improved their cognition and performance on coding, mathematical, and pattern recognition problems (Banderet and Lieberman 1989). Similar results were found in an investigation where adult males were sleep-deprived for 30 h and then given 150 mg/kg body weight dose of tyrosine in capsule form. Despite extended sleep loss, subsequent to the tyrosine supplement, the subjects showed improvements in mathematical processing, as well as their response times for running memory, logical reasoning, and visual vigilance (Magill et al. 2003).

5.12   Effect of Tyrosine on Mood

Dopamine and norepinephrine deficiencies have been implicated in depression; hence, it is not surprising that a number of investigations have focused on the relationship of tyrosine to mood. Tyrosine ingestion has shown to modify the neurotransmission of these catecholamines (Wurtman et al. 1980). Plasma levels of tyrosine are significantly decreased in depressed patients as compared to controls, and plasma levels rise as depressed patients recover. Less tyrosine may be transported to the brain of depressed individuals, and a reduced ratio of plasma tyrosine to the other LNAAs seems to correlate with responsiveness to imipramine, a prescription antidepressant (Spring 1986). In a study that enrolled nondepressed subjects, Leatherwood and Pollet (1983) administered 500 mg tyrosine along with a low-protein, high-carbohydrate meal. No significant effects were shown for mood, but the high-carbohydrate content of the meal likely enhanced tryptophan levels and thereby counteracted any effects of the tyrosine supplement. To improve upon this experimental flaw, Lieberman and his colleagues administered 100 mg/kg body weight of tyrosine to 20 healthy males in a double-blind crossover study, but on an empty stomach following an overnight fast. However, they also found no mood-altering effect of tyrosine (Lieberman et al. 1983).

Since many of the drugs used to treat depression appear to enhance neurotransmission of norepinephrine, the hypothesis that depression results from a deficiency in norepinephrine has been posited. Some early research with a handful of depressed patients showed improvement in symptoms with tyrosine treatment and a relapse when placebo was substituted (Goldberg 1980). However, in none of the case studies was tyrosine administered in a blind manner.

Employing a double-blind experimental design, Gelenberg et al. (1983) assigned outpatients suffering from severe depression to either a tyrosine treatment group (100 mg/kg body weight/day) or to a placebo control condition for a 4-week period. Although 67% of the treatment group improved as compared to 38% of the placebo group, these results are less dramatic when one considers that the tyrosine and placebo groups were compared of just six and eight patients, respectively (Gelenberg et al. 1983). Hence, tyrosine was effective in four out of six patients, while the placebo produced similar effects in three out of eight. Subsequently, Gelenberg et al. (1983) conducted a larger trial, randomly assigning 65 depressed outpatients to tyrosine (100 mg/kg body weight/day), imipramine (2.5 mg/kg body weight/day), or placebo for a minimum of 4 weeks. Although catecholamine production was enhanced, the tyrosine group was no different than the placebo group at the end of the treatment (Gelenberg et al. 1990). In fact, all three groups did improve, but the imipramine group showed a trend toward the greatest improvement. These results seem to provide modest support for the efficacy of tyrosine in elevating mood or treating depression, although a population-based investigation in Finland found no association between dietary intake of tyrosine and self-reports of depressed mood (Hakkarainen et al. 2003). However, research in this area will most likely continue.

References

Abbot, N. J., Ronnback, L., & Hansson, E. (2006). Astrocyte-endothelial interactions at the blood-brain barrier. *Nature Reviews Neuroscience, 7,* 41–53.

Amer, A., Breu, J., McDermott, J., Wurtman, R. J., & Maher, T. J. (2004). 5-hydroxy-L-tryptophan suppresses food intake in food-deprived and stressed rats. *Pharmacology, Biochemistry & Behavior, 77*(1), 137–143.

Aquilonius, S., & Eckeras, S. (1977). Choline therapy in Huntington's chorea. *Neurology, 27,* 887–889.

Banderet, L. E., & Lieberman, H. R. (1989). Treatment with tyrosine, a neurotransmitter precursor; reduces environmental stress in humans. *Brain Research Bulletin, 22,* 759–762.

Blundell, J. E., & Hill, A. J. (1987). Influence of tryptophan on appetite and food selection in man. In S. Kaufmann (Ed.), *Amino acids in health and disease: New perspectives* (pp. 403–419). New York: Alan R, Liss.

Charles, H. C., Lazeyras, F., Krishnan, K. R. R., Boyko, O. B., Payne, M., & Moore, D. (1994). Brain choline in depression: In vivo detection of potential pharmacodynamic effects of antidepressant therapy using hydrogen localized spectroscopy. *Progress in Neuropsychopharmacology & Biological Psychiatry, 118,* 1121–1127.

Chew, M. L., Mulsant, B. H., Pollock, B. G., Lehman, M. E., & Greenspan, A. (2008). Anticholinergic activity of 107 medications commonly used by older adults. *Journal of the American Geriatrics Society, 56,* 1333–1341.

Christensen, L. (1996). *Diet-behavior relationships: Focus on depression.* Washington, DC: American Psychological Association.

Corell, C. U., Leucht, S., & Kane, J. (2004). Lower risk for tardive dyskinesia associated with second-generation anti-psycholics: A systematic review of 1-year studies. *American Journal of Psychiatry, 161,* 414–425.

Corsica, J. A., & Spring, B. J. (2008). Carbohydrate craving: A double-blind, placebo-controlled test of the self medication hypothesis. *Eating Behaviors, 9,* 447–454.

Davidson, M., Stern, R. G., Bierer, L. M., Horvath, T. B., Zemishlani, Z., Markofsky, R., & Mohs, C. S. (1991). Cholinergic strategies in the treatment of Alzheimer's disease. *Acta Psychiatrica Scandinavica, 366,* 47–51.

Delgado, P. L., Price, L. H., Miller, H. L., Salomon, R. M., Licinio, J., Krystal, J. H., Heninger, G. R., & Charney, D. S. (1991). Rapid serotonin depletion as a provocative challenge test for patients with major depression: Relevance to antidepressant action and the neurobiology of depression. *Psychopharmacology Bulletin, 27,* 321–329.

Ellenbogen, M. A., Young, S. N., Dean, P., Palmour, R. M., & Benkelfat, C. (1996). Mood response to acute tryptophan depletion in healthy volunteerts: Sex differences and temporal stability. *Neuropharmacology, 15*(5), 465–474.

Ereshefsky, L., Rospod, R., & Jann, M. (1989). Organic brain syndromes, Alzheimer type. In J. T. DiPiro, R. L. Talbert, P. E. Hayes, G. C. Yee, & L. M. Posey (Eds.), *Pharmacotherapy: A pathophysiological approach* (pp. 678–696). New York: Elsevier.

Fernstrom, J. D. (2000). Can nutrient supplements modify brain function? *American Journal of Nutrition, 7,* 1669S–1673S.

Gelenberg, A. J., Wojcik, J. D., Gibson, C. J., & Wurtman, R. J. (1983). Tyrosine for depression. *Journal of Psychiatric Research, 17,* 175–180.

Gelenberg, A. J., Wojcik, J., Falk, W. E., Bellinghausen, B., & Joseph, A. B. (1989). CDP-choline for the treatment of tardive dyskinesia: A small negative series. *Comprehensive Psychiatry, 30,* 1–4.

Gelenberg, A. J., Wojcik, J., Falk, W. E., Baldessarini, R. J., Zeisel, S. H., Schoenfeld, D., & Mok, G. S. (1990). Tyrosine for depression: A double-blind trial. *Journal of Affective Disorders, 19,* 125–132.

Goldberg, I. K. (1980). L-tyrosine in depression. *The Lancet, 2,* 364.

Growdon, J. H., Hirsch, M. J., Wurtman, R. J., & Wiener, W. (1977). Oral choline administration to patients with tardive dyskinesia. *New England Journal of Medicine, 297*, 524–527.

Hakkarainen, R., Partonen, T., Haukka, J., Virtamo, J., Albanes, D., & Lonnqvist, J. (2003). Association of dietary amino acids with low mood. *Depression and Anxiety, 18*, 89–94.

Hirsch, M. J., & Wurtman, R. J. (1978). Lecithin consumption elevates acetylcholine concentration in rat brain and adrenal gland. *Science, 202*, 223–225.

Hudson, C., Hudson, S. P., Hecht, T., & Mackenzie, J. (2005). Protein source tryptophan versus pharmaceutical grade tryptophan as an efficacious treatment for chronic insomnia. *Nutritional Neuroscience, 8*, 121–127.

Jackson, I., Nuttall, A., & Perez-Cruet, J. (1979). Treatment of tardive dyskinesia with lecithin. *American Journal of Psychiatry, 136*, 1458–1459.

Janowsky, D. S., & Overstreet, D. H. (1998). Acetylcholine. In P. J. Goodnick (Ed.), *Mania: Clinical and research perspectives* (pp. 135–155). Washington, DC: American Psychiatric Press, Inc.

Jouvet, M. (1968). Insomnia and decrease of cerebral 5-hydroxytryptamine after destruction of the raphe system in the cat. *Advances in Pharmacology, 6*, 265–279.

Kanarek, R. B. (1987). Neuropharmacological approaches to studying diet selection. In S. Kaufman (Ed.), *Amino acids in health and disease: New perspectives* (pp. 383–401). New York: Alan R. Liss.

Kompoliti, K., & Horn, S. S. (2007). Drug-induced and iatrogenic neurological disorders. In C. G. Goetz (Ed.), *Textbook of clinical neurology* (pp. 1285–1318). Philadelphia: Saunders Elsevier.

Leatherwood, P. D., & Pollet, P. (1983). Diet-induced mood changes in normal populations. *Journal of Psychiatric Research, 17*(2), 147–154.

Leibenluft, E., Fiero, P., Bartko, J. J., Moul, D. E., & Rosenthal, N. E. (1993). Depressive symptoms and the self-reported use of alcohol, caffeine, and carbohydrates in normal volunteers and four groups of psychiatric outpatients. *American Journal of Psychiatry, 150*, 294–301.

Lieberman, H. R., Corkin, S., Spring, B. J., Growdon, J. H., & Wurtman, R. J. (1983). Mood, performance, and pain sensitivity: Changes induced by food constituents. *Journal of Psychiatric Research, 1983*, 135–145.

Lopez-Figueroa, A. L., Norton, C. S., Lopez-Figueroa, M. O., Armelini-Dodel, D., & Burke, S. (2004). Serotonin 5-HT1A, 5-HT1B, and 5-HT2A receptor mRNA expression in subjects with major depression, bipolar disorder and schizophrenia. *Biological Psychiatry, 55*, 225–233.

Lucca, A., Lucini, V., Piatti, E., Ronchi, P., & Smeraldi, E. (1992). Plasma tryptophan levels and plasma tryptophan/neutral amino acids ratio in patients with mood disorder, patients with obsessive-compulsive disorder and normal subjects. *Psychiatric Research, 44*, 85–91.

Magill, R. A., Waters, W. F., Bray, G. A., Volaufova, J., & Smith, S. R. (2003). Effects of tyrosine, phentermine, caffeine D-amphetamine, and placebo on cognitive and motor performance deficits during sleep deprivation. *Nutritional Neurosciences, 6*, 237–246.

Mayeno, A. N., & Gleich, G. J. (1994). Eosinophilia-myalgia syndrome and tryptophan production: A cautionary tale. *Trends in Biotechnology, 12*, 346–352.

Menkes, D. B., Coates, D. C., & Fawcett, J. P. (1994). Acute tryptophan depletion aggravates premenstrual syndrome. *Journal of Affective Disorders, 32*(1), 37–44.

Nathan, P. J., Baker, A., Carr, E., Earle, J., Jones, M., Nieciecki, M., Hutchison, C., & Stough, C. (2001). Cholinergic modulation of cognitive function in healthy subjects: Acute effects of donepezil, a cholinesterase inhibitor. *Human Psychopharmacology, 16*(6), 481–483.

Nicholas, D. E., & Nicholas, C. D. (2008). Serotonin receptors. *Chemical Reviews, 108*, 1614–1641.

Owasoyo, J. O., Neri, D. F., & Lamberth, J. G. (1992). Tyrosine and its potential use as a countermeasure to performance decrement in military sustained operations. *Aerospace Medical Association, 63*, 364–369.

Pardridge, W. M. (1986). Blood-brain barrier transport of nutrients. *Nutrition reviews, 44*, 15–25.

Petty, F., Kramer, G., Wilson, L., & Chae, Y. L. (1993). Learned helplessness and *in vivo* hippocampal norepinephrine release. *Pharmacology, Biochemistry and Behavior, 46*, 231–235.

Rolls, B. J., Shide, D. J., Thorwart, M. L., & Ulbrecht, J. S. (1998). Sibutramine reduces food intake in non-dieting women with obesity. *Obesity Research, 6*, 1–11.

Rosenberg, G. S., & Davis, K. L. (1982). The use of cholinergic precursors in neuropsychiatric diseases. *American Journal of Clinical Nutrition, 36*, 709–720.

Ruhe, H. G., Mason, N. S., & Schene, A. H. (2007). Mood is indirectly related to serotonin, norepinephrine and dopamine levels in humans: A meta-analysis of monamine depletion studies. *Molecular Psychiatry, 12*, 331–359.

Schwartz, J. H. (2001). *Neurotransmitters. Encyclopedia of life sciences*. Wiley Online Library. https://doi.org/10.1038/npg.els.0000287.

Sitaram, N., Weingartner, H., Caine, E. D., & Gillin, J. C. (1978). Choline: Selective enhancement of serial learning and encoding of low imagery words in man. *Life Sciences, 22*, 1555–1560.

Smith, M. J., & Garrett, R. H. (2005). A heretofore undisclosed crux of eosinophilia-myalgia syndrome: Compromised histamine degradation. *Inflammation Research, 54*, 435–450.

Spring, B. (1986). Effects of foods and nutrients on the behavior of normal individuals. In R. J. Wurtman & J. J. Wurtman (Eds.), *Nutrition and the brain* (pp. 1–48). New York: Raven Press.

Spring, B., Chiodo, J., Harden, M., Bourgeois, M. J., Mason, J. D., & Lutherer, L. (1989). Psychobiological effects of carbohydrates. *Journal of Clinical Psychiatry, 50*(5), 27–33.

Sved, A. F. (1983). Precursor control of the function of monoaminergic neurons. In R. J. Wurtman & J. J. Wurtman (Eds.), *Nutrition and the brain* (Vol. 6, pp. 223–275). New York: Raven Press.

Wecker, L. (1990). Choline utilization by central cholinergic neurons. In R. J. Wurtman & J. J. Wurtman (Eds.), *Nutrition and the brain* (Vol. 8, pp. 147–162). New York: Raven Press.

Wurtman, R. J., & Wurtman, J. J. (1995). Brain serotonin, carbohydrate craving, obesity and depression. *Obesity Research, 3*, 477S–480S.

Wurtman, R. J., Hirsch, M. J., & Growdon, J. H. (1977). Lecithin consumption raises serum free choline levels. *Lancet, 2*, 68–69.

Wurtman, R. J., Hefti, F., & Melamed, E. (1980). Precursor control of neurotransmitter synthesis. *Pharmacological Reviews, 32*, 315–335.

Yokogoshi, H., & Wurtman, R. J. (1986). Metal composition and plasma amino acid ratios: Effects of various proteins on carbohydrates, and of various protein concentrations. *Metabolism, 35*, 837–842.

Young, S. N. (1991). Some effects of dietary components (amino acids, carbohydrate, folic acid) on brain serotonin synthesis, mood, and behavior. *Canadian Journal of Pharmacology, 69*, 893–903.

Young, S. N. (2005). Amino acids, brain metabolism, mood and behavior. In H. R. Lieberman, R. B. Kanarek, & C. Prasad (Eds.), *Nutritional neuroscience* (pp. 131–146). Boca Raton: Taylor and Francis.

Young, S. N., Smith, S. E., Pihl, R. O., & Ervin, F. R. (1985). Tryptophan depletion causes a rapid lowering of mood in normal males. *Psychopharmacology, 87*, 173–177.

Yuwiler, A., Oldendorf, W. H., Geller, E., & Braun, L. (1977). Effect of albumin binding and amino acid competition on tryptophan uptake into brain. *Journal of Neurochemistry, 28*, 1015–1023.

Chapter 6
Undernutrition: The Bane of Modern Times and It's Developmental Fallout

Abstract The chapter discusses, principally, undernutrition and its developmental fallout. Emphasis is placed on Protein-Energy -Malnutrition (PEM), with the primary focus on the African context.

Keywords Malnutrition and its ecology · Protein-Energy-Malnutrition (PEM) · Human development · Infants · Cognitive deficits · Behavioral problems · Academic performance

The term "malnutrition" has been so thoughtlessly used in scientific literature and developmental investigations. This is more so with international "aid-giving" agencies, which, more often than not, when applied to poor countries in the developing region of Africa, Asia, and Latin America, have a hidden agenda. Is a scrawny child with a bloated belly in a remote corner of Africa malnourished or an urban child, living in a posh apartment in a city in a western country, who goes to bed each night hungry also malnourished? Or, perhaps, prosaically put, an 11-year-old girl, who has started to frenetically "diet," after she first menstruated, malnourished? All these examples comprise malnutrition, if we take it to mean a condition caused by deficient nutrient intake. But, even an obese individual may be considered malnourished, if we take it to mean that his or her calorific consumption is far in excess of what he or she needs to maintain a healthier weight and lifestyle.

In this chapter, malnutrition will be discussed in the traditional sense of the word, the condition caused by acute or prolonged undernutrition. While those who are fortunate to live in developed countries may only be acquainted with severe undernutrition from watching television, protein-energy malnutrition (PEM) continues to be a major problem in the world. Also referred to as protein-calorie malnutrition, PEM has long been recognized as a consequence of poverty, but it has become increasingly clear that it is also a cause of poverty (UNICEF 1998). Having been robbed of their mental and physical potential, children who manage to survive PEM and live into adulthood have lessened intellectual ability, lower levels of productivity, and higher incidences of chronic illness and disability.

© Springer Nature Switzerland AG 2020
K. P. Nair, *Food and Human Responses*,
https://doi.org/10.1007/978-3-030-35437-4_6

Although malnutrition is associated with poverty in the third world, recent reports have clearly demonstrated that modern societies, including those in the USA and Europe, are not immune to this problem. In the USA, it is estimated that more than one in every four children under age 12 do not get all the food they need, while researchers in the UK have documented the health risks linked to diet in poor families. Among low-income populations in these developed societies, low birth weight, anemia, growth failure, weakened resistance to infection, increased susceptibility to lead poisoning, dental disease, and so on, which are all associated with undernutrition and, unfortunately, are not all that infrequent (James et al. 1997).

6.1   Contours of Protein-Energy Malnutrition

The term "malnutrition" is a wide canvas, of which, PEM is just one. Besides PEM, deficiencies in iron, iodine, and vitamin A may appear in combination and contribute to the debilitating effects of each other. Chronic PEM is the most common form of malnutrition in the world, affecting an estimated 171 million children (Save the Children 2012). Of the 7.6 million children under age 5, who die each year in developing countries from preventable causes, the deaths of 2.6 million or 35% are either directly or indirectly attributable to malnutrition – due to its association with diarrhea, malaria, measles, respiratory infections, and prenatal problems (Rice et al. 2000). Adults also suffer from hunger and starvation; however, PEM occurs most frequently and also has its most devastating consequences in infancy and early childhood.

Severe PEM in infants and children has traditionally been classified either as *marasmus* or as *kwashiorkor*. The first results from insufficient energy intake, that is, extremely low intake of both protein and calories. The word derives from the Greek root *marasmus*, which means "to waste away." In this context, wasting refers to the wasting away of fat and muscle, with a much lower weight than would be expected, given the child's height. By contrast, stunting refers to the impact of PEM on height, with the child being considerably shorter than he or she should be, given his or her age.

Marasmus is the most often observed disease in infants under 1 year of age at the time of weaning (Fig. 6.1).

In developed countries a form of this can strike young women who have dieted excessively, as anorexia, as well as the elderly poor who have difficulty in obtaining or ingesting sufficient calories. By contrast, *kwashiorkor* results from insufficient intake of protein, with caloric needs usually satisfied. The word itself literally means "the disease of the deposed baby when the next one is born" and derives from the language of the Ga tribe of Ghana. As this meaning suggests, kwashiorkor is most likely to occur in the second or third year of life, when a baby is weaned to make way for the next one born (Fig. 6.2).

The majority of severely malnourished children in underdeveloped nations often exhibit symptoms of both conditions or alternately display one and then the other,

Fig. 6.1 Marasmic infant.
(Photo credit: Collection
Getty Images)

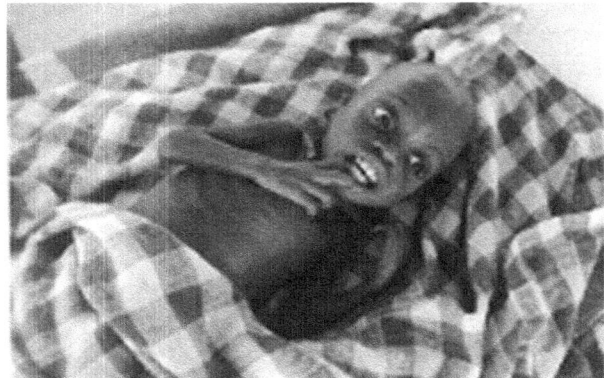

Fig. 6.2 Infant with
kwashiorkor. (Photo credit:
Collection Getty Images)

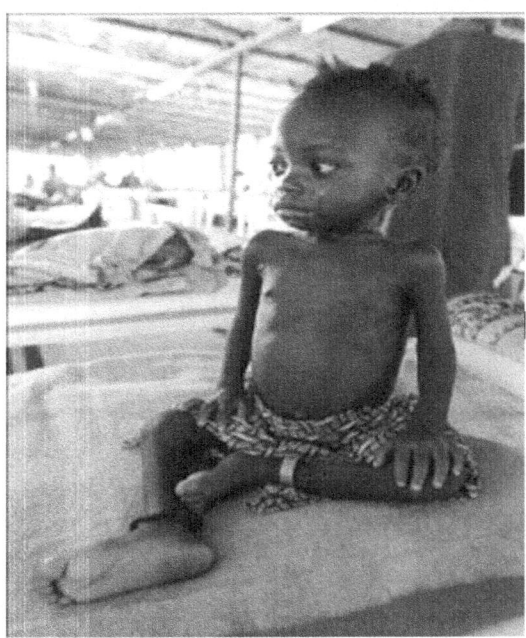

and this is referred to as *marasmic kwashiorkor.* An adult form of malnutrition, known as *iatrogenic PEM*, characterizes the individual whose nutritional status deteriorates after prolonged hospitalization, when hospitalization is not due to a dietary or gastrointestinal problem.

Marasmus is the form of PEM most commonly affecting infants in the age group of 6–18 months, when they are weaned from breast milk to formula feed. In attempting to stretch their limited resources, the impoverished mother will dilute the commercial formula, further contaminating it with dirty water, which leads to bacterial infection. The infant then suffers from repeated episodes of gastroenteritis, resulting in diarrhea, dehydration, and anemia. In an attempt to cure the baby, the mother may

Table 6.1 A comparison of various forms of PEM

Feature	Marasmus	Kwashiorkor	Iatrogenic
Type of victim	Infants, elderly poor	Toddlers	Hospitalized adults
Growth failure	Severe	Somewhat severe	Not applicable
Muscle changes	Wasting	Wasting	Weakness
Subcutaneous fat	Absent	Present	Decreased
Edema	Absent	Always	Absent
Skin changes	Rare	Frequent	Capillary fragility
Hair changes	Frequent	Very frequent	Frequent
Liver enlargement	Rare	Frequent	Rare
Diarrhea	Frequent	Frequent	Frequent
Blood changes	Frequently anemic	Anemia	Low lymphocyte count
Serum albumin	Normal	Low	Low
Appetite	Ravenous	Anorectic	Anorectic
Irritability	Always	Always	Frequent
Apathy	Always	Always	Apathy toward eating
Other psychological	Failure to thrive	Whimpering cry	Altered taste sensation

even withhold food. Recurrent infections, coupled with little or no food, subsequently lead to marasmus. In terms of resultant behavior, the marasmic infant is typically irritable but may also appear weak and apathetic. The infant is also understandably hungry but can seldom tolerate large amounts of food and will easily vomit (Torun and Viteri 1988). The symptoms of the different conditions are given in the following table (Table 6.1).

Kwashiorkor, which typically develops when the child is 18–24 months old, comes about when a new born baby arrives, and the toddler is weaned from the breast and fed the high-carbohydrate-low-protein diet that the rest of the family subsists on. While this diet may be adequate for the adult, it is insufficient for the rapidly growing toddler. Since the child of this age begins to explore the environment, the opportunity for exposure to bacterial and viral infection can further exacerbate the consequences of the poor diet. However, one of the behavioral characteristics of this is a lessened interest in the environment, along with irritability, apathy, and frequent anorexia. But most notably, the toddler with kwashiorkor cries easily and often displays an expression of sadness and misery (Torun and Viteri 1988).

Iatrogenic PEM occurs most commonly in patients, often seniors, who are hospitalized longer than 2 weeks, and can result from acts and omissions by healthcare providers or from institutional policies and practices which undermine optimal nutritional care. Patients may report a lack of interest in, an aversion toward or a perceived inability to eat, along with altered taste sensations. Reduced eating combined with extended time in bed leads to poor muscle tone (including those muscles required for chewing) and swallowing and mental irritability or confusion.

6.2   PEM Juxtaposed with Early Human Development

It is a fallacy to think that chronic poverty is associated only with large families, as it has been found that severe malnutrition and food deprivation can reduce fertility (Gopalan and Naidu 1972). It is important to note that from an anthropological and behavioral perspective, both men and women may have little interest in sex during extreme times of starvation. Biologically speaking, women may develop amenorrhea, while men may lose their ability to produce viable sperm. From an anthropological perspective, this might be nature's way of controlling the human population growth corroborating the common axiom "struggle for existence and survival of the fittest." If a malnourished woman becomes pregnant, she will have to face the challenge of supporting both the growth of her fetus and her own physical health with less than sufficient nutrient uptake. Half a century ago, it was believed that through a protective, almost parasitic relationship with the mother, the fetus would be spared any adverse consequences of maternal malnutrition (Naismith 1969). There are, in fact, two mechanisms that buffer the fetus from adequate nutrition. First of all, the mother's intake of food provides a direct nutritional source. But in addition, the placenta transfers nutrients which are stored by the mother to the developing fetus. However, malnutrition before or around the time of conception can prevent the placenta from developing fully.

Severe PEM early in development results typically in failure to maintain embryonic implantation, resulting in spontaneous abortion. Moderate malnutrition throughout gestation will generally permit the continued development of the fetus but will also lead to changes in the growth of both the placenta and the fetus. If the placenta is poorly developed, it cannot deliver proper nourishment to the fetus, and the infant may subsequently be born prematurely, small for gestational age, and with reduced head circumference. In human infants, a reduced head circumference may be the first indication that malnutrition, in particular during gestation and infancy, leads to permanent brain damage. Since *marasmus* typically develops at a younger age than *kwashiorkor*, the marasmic child is more likely to have a reduced head circumference than one with *kwashiorkor*.

Indeed, enormous scientific interest has been seen to take place in understanding the effect of malnutrition on the developing fetus. A current topic of widespread interest, referred to as the *Barker hypothesis* (alternatively the "thrifty phenotype," the "fetal programming," or the "developmental origin of adult disease" hypothesis), proposes that a significant number of adulthood diseases may trace their origins to undernutrition during fetal development (Barker 1992). Barker and his colleagues have shown that during prenatal development, the fetus responds to severe malnutrition by favoring the metabolic demands of the growing brain and central nervous system, as well as the heart, *at the expense of the other tissues*. In other words, in an intrauterine environment which lacks the nutrients which the infant needs, for proper organ growth, the fetus may react by slowing its metabolism to conserve all the energy-rich fat that it can. This, in turn, may lead to obesity, heart disease, hypertension, and diabetes in later life. Alternatively, the fetus may favor the brain over an organ like the liver, for example, when it must subsist on inadequate sources of energy. Despite some persuasive epidemiological evidence for the hypothesis, the scientific and medical communities did not initially embrace the concept. However, scores of investigators are now attempting to test this "fetal programming" hypothesis, and the coming years should see further evidence that bears on this important question, with implications for its prevention, as well (Dover 2009).

On the whole, investigations related to malnutrition in infants and its consequences are rather limited. Over four decades, Winick and Rosso (1969) observed significant reductions in brain weight, total protein, and total DNA content in infants who died of *marasmus* before they were a year old. Those infants with extremely low birth weights, indicative of prenatal malnutrition, had lower brain DNA content than did the malnourished infants with higher brain weights. The reduced DNA content is particularly meaningful, as it indicates a decrease in both neurons and glial cells.

Future efforts to determine the effects of malnutrition on brain development will likely rely on the availability of magnetic resonance imaging (MRI), which can even be done on the fetus before birth. Reports on MRI have confirmed that brain myelination is impaired in malnourished infants (Georgieff 2007).

Evidence concerning the role of malnutrition on development of postnatal brain has generally supported the hypothesis that development is most impaired in those cell types and regions which show maximum growth at the time that nutritional deficiency is present. Hence, postnatal malnutrition is usually not associated with a reduction of neuron number but in glial cell number. However, the primary effect of postnatal malnutrition is a reduction in the size of both neurons and glial cells, with synaptogenesis and myelination possibly being inhibited.

Since kwashiorkor develops later in postnatal life, it has a less permanent effect on brain development than *marasmus*. The disease does not lead to marked lessening in neuron number. An early investigation of infants who died during their second year of life showed only minor deficits in brain DNA level, though the decreased brain weight\DNA ratio indicated a reduction in brain cell size. Providing adequate nutrition in later life led to some degree of recovery (Super et al. 1990). In one MRI investigation which monitored changes in cerebral atrophy, it was shown that brain shrinkage accompanying *kwashiorkor* was reversed through nutritional rehabilitation in later life.

6.3   Severe Malnutrition: How Does It Affect Human Behavior?

The last four decades have seen a remarkable understanding of the relationship of PEM to behavioral development, in particular intellectual competence. The focal question was whether malnutrition will also lead to lessened cognitive competence, due to impaired brain development. Both underperformance and lower IQ scores were reported in malnourished children. Based on such findings, policy makers, and well-meaning scientists as well, concluded that malnutrition in children was a direct cause of impaired mental development owing to its adverse effects on the brain development. A straightforward explanatory model was endorsed as follows:

Malnutrition —— Brain Damage —— Impaired Behavior with lower IQ

A corollary to this model was that improving dietary management would lead to a significant enhancement of the subject's intellectual capacity (Ricciuti 1993). It is now generally acknowledged that this model, though simplistic, falls rather short of explaining how malnutrition compromises mental development (Gorman 1995). Rather, a myriad of adverse health and socio-environmental conditions are now seen as interacting with nutritional status to influence mental development, as well as behavioral outcomes. This is the central theme on which "supposedly philanthropic" agencies, in particular in the West, target the poorer regions of the globe, forgetting the important fact that "trickle down development" never works. This is the tragedy Africa experienced and even now continues to experience. The same is true of many areas on the Asian and Latin American continents. Leaving aside the "politics of aid," it might be well worth to discuss some key issues, as expounded in the following paragraphs.

6.4   Malnutrition: It's Ecology

The starting point on any aspect of malnutrition discussion should be that, at the center of this human tragedy, it is the lack of protein and calories which lead to PEM. But, the problem of PEM cannot be viewed in isolation from other socio-environmental conditions (Wachs 1995). Indeed, the environment of a malnourished child is different in countless ways, beyond the mere deficiencies of food intake. In this context, it is worth noting that the UN Food and Agriculture Organization, headquartered in Rome, has been conducting many "Food Summits," attended by many heads of States from all over, and the "entrenched food, agriculture, and developmental scientists." This author, as an "outsider" of the entrenched system, once had the temerity to publicly suggest that Food Summits cannot be discussed on a full belly – it has been the tradition to arrange lavish lunches and dinners, coupled with a 5-star hotel stay in Rome, of course at enormous expense to the United Nations, for the delegates, and meet other incidental expenses. Instead, people who sermonize at these summits should really know what it means to be hungry. The current US President, Donald Trump, had the courage and frankness to suggest, during his election campaign, that those who draw their salaries on the UN budget, to which the USA is a significant contributor "have a good time but do nothing worthwhile." The same goes with these food summits. Only when one is hungry, would he or she realize the pangs of hunger and what it means to be truly hungry and the importance of adequate intake of enough food. It used to be a spectacle seeing delegates from the Asian and African continents expounding sermons on what should be done on the "food front," while many of their own countrymen, back home, were starving to death. And India, with its so-called green revolution, a borrowed idea from the USA, which was nothing but a highly soil extractive and soil exploitative farming, leaving countless environmental hazards in its wake, claiming to have brought about "food self-sufficiency," has its own tale of hunger. This author has persistently taken the position that if, indeed, India is "self-sufficient" in food production, why is food today (2017 as of writing this book), so expensive, in India, with rural households spending as much or more than 50% of their daily income on food? Go to China, it has a different story to tell. With a population at close to 1.5 billion (India's own is now touching 1.3 billion), it has harvested last year more than 500 million tons of food grain and India with its less than 100 million tons production, pales in comparison. And, to boot, we now have the same scientists who ushered the so-called green revolution, now telling Indian farmers to go in for a "second green revolution" with genetically modified crops, which are shrouded in scientific controversies. We had these scientists telling us that the green revolution was needed because of the Bengal famine – a white lie, indeed. In fact, the Bengal famine was the outcome of the British exporting food produced on Indian soil to feed the military during the World War II. A case of merciless exploitation of the exploited.

This author chose to dwell on these abovementioned aspects of food, only in passing, as it is not the principal topic of this book, but a truly related one, as the current discussion in this chapter is on limited food intake and its fallout on the health of adults and children.

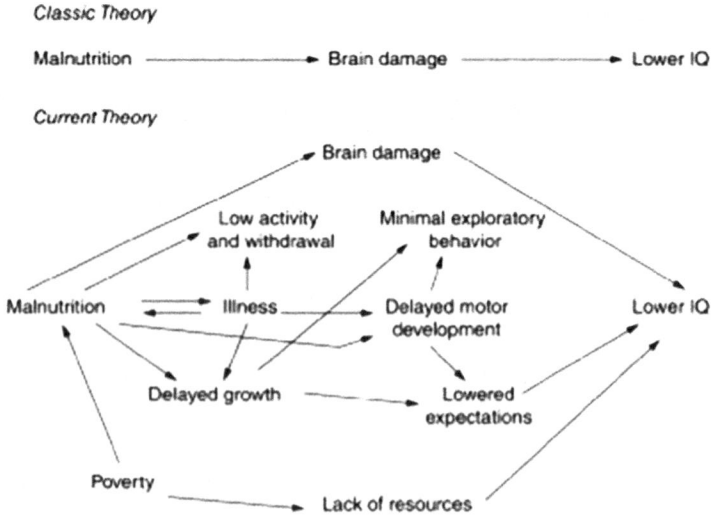

Fig. 6.3 Effect of malnutrition on cognitive development. (*Source:* Brown and Pollitt 1996)

The following figure (Fig. 6.3) displays the numerous forces which are likely to interact to reduce intellectual development. To be sure, the brain may be directly affected by PEM, but this model allows for a reversal if adequate nutrition is made available. Health may also be compromised, but illness itself can further reduce the child's level of nutrient intake. From an ecological perspective, however, it is the reduction in the child's energy level, slowed motor development, minimal exploration of the environment, and lowered expectation by parents that may place the impoverished child at greater risk of impaired cognitive development (Brown and Pollitt 1996).

The impact of poverty cannot be underestimated when discussing the effect of PEM, because a child's state of malnourishment is invariably confounded with socioeconomic status. The term *confound* here means that poor socioeconomic status co-varies with poor nutrition, and since both of these factors may be independently associated with reduced cognitive performance in children, it is difficult to determine if one, the other, or their interaction is most responsible for mental defects. From what has been discussed so far, the hypothesized link from malnutrition to impaired mental development should be obvious. However, the link from lower socioeconomic status to child development merits further explanation.

Less money available for food means less food on the table. Less food also means less variety of foods served and conceivably less interest in eating. This is a vicious cycle among the poor. One of the notable aspects of high costs of the available food, for the poor, notwithstanding the so-called Public Distribution System, as, for example, in a country like India – where food meant for the poor is siphoned off through the unscrupulous alliance of corrupt politicians and merchants of food, to deprive the really needy and divert the food from the government kitty (the Food Corporation of India) to private marketing – is that for these poor, apart from the basic staples like rice or wheat, supplementary availability of fruits, fish, eggs, poultry, and meat are out of reach for the economically deprived, as these are expensive items of food in India. Thus, for these poor, eating becomes "less attractive" as there is no variety of food available on the plate. This is a silent but sorrowful consequence to the poor, which, in turn, adversely affects the cognitive capacity and behavioral aspects of those involved.

Children learn much from their experience with food, in terms of tastes that stimulate their experiences with food, because of gustatory and olfactory senses. Early food experiences also teach children labels, amounts, and physical processes inherent in preparing and cooking different food items (Endres et al. 2004). Less money available also means fewer resources in general, for clothing and shelter. When one is trying to survive, play is a luxury that poor families can ill afford. Not only is the opportunity for play restricted, but less money also means, few if any, toys that would serve to stimulate cognitive development. Parents in poor villages or neighborhoods may not view the area as safe, so they further prevent their children from freely exploring the local environs. Thus, family economic background and parental education are known to positively influence children's cognitive development (Hackman and Farah 2009). Parents of malnourished children are likely to be malnourished themselves, further reducing their interest in making time to stimulate their offspring. It is worth noting that in the only published prospective study of children which included observations of mother-child days *before* the children were malnourished, mothers were found to be less responsive and affectionate to their children and spoke less to them as well (Cravioto 1977). Indeed, one may speculate that the strongest predictions of developmental outcomes can be made when based on the interaction of nutrient status with such contextual factors as low socioeconomic status, low parental education, family illiteracy, and specific types of caregiver-child transactions (Wachs 1995).

6.5   Effects in Infants

Malnutrition-cognitive development interaction has been well researched. While tests of normative abilities are better at gauging performance relative to peers of the same age than in estimating actual intelligence, a number of tools such as Bayley Scales of Infant Development (please refer chap. 4) or the Griffith Mental Development Scales have been used in studies of malnourished infants. Regardless of the type of malnutrition, the development of infants and toddlers on such tests is significantly delayed (Grantham-McGregor 1995). In general, language and verbal development appear to be the most negatively affected by PEM. Malnutrition has also been associated with abnormalities in cerebral functioning as measured by psychophysiological tests of arousal. For example, heart rate deceleration is expected to occur when an infant is presented with a novel stimulus, with habituation occurring when the infant becomes habituated to the stimulus. As it turns out, this effect may only be seen with well-nourished infants. When malnourished infants viewed a novel stimulus, they displayed no alterations in heart rate (Lester et al. 1975). The infants' ability to integrate intersensory information, that is, to tell that a particular sound and image go together, is also delayed in malnourished infants (Cravioto and Arrieta 1986). These data suggest that severe malnutrition leads to a diminished responsiveness to environmental signals and therefore to a reduction in the infant's capacity for information processing. It is likely that these defects in information processing will persist even after nutritional rehabilitation has occurred.

Of all the behavioral symptoms associated with marasmus and kwashiorkor, lethargy and reduced activity are the most commonly observed. In fact, it has been hypothesized that this reduction in motor activity may serve to isolate malnourished infants from their environment, resulting in limited opportunities for learning and thereby depression and mental development (Schurch and Scrimshaw 1990). Specifically, the *functional isolation* hypothesis posits that the decreased activity, which characterizes malnourished infants, makes them less likely to seek stimulation from their environment (Brown and Pollitt 1996). In turn, the caregivers are less responsive to their infants and offer them less stimulation. Over time, this repeating circle of less seeking/less receiving of stimulation interferes with the infant's normal acquisition of information from social and physical environment. Left unchecked, the child's development is adversely affected (Wachs 2009).

Besides the reduction in motor activity which characterizes PEM, there are other non-cognitive aspects of the malnourished infant's behavior which bear on the interaction between mother and the baby and further exacerbate the risk of lowered stimulation. An infant who shows little social responsiveness to his mother, and who persists in fussing and crying when awake, cannot be much fun to be around. Add to this scenario the fact that the mother is likely to be malnourished, as well. With less energy to stimulate, rouse up, or play with her baby, her caregiving efforts are likely to be minimal, and the infant's cognitive development is bound to suffer. Such a cycle of depressed mother-infant interaction is sometimes seen in cases of *nonorganic failure to thrive*. Even in the USA with developed societies, some 1–5% of infants who by all accounts should be well nourished nevertheless can show deterioration in their rate of growth. Although some cases of failure to thrive are classified as organic in origin, the more common, nonorganic type consists of abnormal behavior and distorted caregiver-infant interactions, in association with the retarded weight gain (Krugman and Dubowitz 2003). Although apathy and reduced activity are characteristic of malnourished infants, many of the behavioral patterns of infants failing to thrive cannot be attributed merely to malnutrition. Instead, the infant's irritability may discourage social interactions, which the mother may interpret as a personal rejection. In other cases, the mother may not be very good at recognizing hunger signals. Alternatively, the mother may recognize the signals, but, because she is under stress or frustrated with her unresponsive baby, she may be less likely to respond herself (Ricciuti 1993).

6.6 What Are the Cognitive Deficits?

In comparison to their non-malnourished siblings, twice as many of the Mexican children who experienced severe malnourishment before 3 years of age had IQs below 70, even after 2 or more years of recovery. Since much of the content of intelligence tests for children consists of language and verbal items, the impact of malnutrition in these areas can directly reduce IQ scores. Unfortunately, malnutrition may have been confounded with poor parenting, as the mothers were less sensitive, verbally communicative, emotionally involved, or interested in their child's performance relative to their behavior with the unaffected child. Galler and Ramsey (1985) conducted a meticulous study of the cognitive consequences of malnutrition in Barbados using a quasi-experimental design known as case control approach, where the investigators carefully matched malnourished children, who had suffered from *marasmus* in infancy, with children of similar social and economic backgrounds but with no history of malnutrition. A retrospective analysis revealed that fathers of the malnourished children held less skilled jobs, and the mothers of those children had less education, were less sociable, and were more depressed than the parents of the matched controls.

An extremely ambitious investigation of malnutrition and its effects on cognitive abilities was conducted in Guatemala, consisting of an early supplementary feeding intervention in four rural villages carefully selected for the high prevalence of PEM (Townsend et al. 1982). Using a placebo-controlled design, two of the villages were randomly assigned to receive a high-protein/high-calorie supplement (known as *atole*) with the other two villages receiving a low-calorie supplement (known as *fresco*). The *atole* supplement was developed by the investigators to serve as a high-protein substitute for the traditional corn gruel that Guatemalan mothers often feed their children, which contained less than 11.5 g protein and 163 kilo calories of energy per cup. The *fresco* drink, common to the villagers, served as the control supplement and contained 59 kilocalories per cup with no protein. Both drinks were fortified with vitamins and minerals. Supplements were administered twice daily ad libitum to all pregnant women, infants, and children 7 years of age and younger, with data collected on over 2300 subjects. The primary hypothesis of the investigation concerned the effects of high-protein/high-energy supplement (*atole*) on cognitive development. The cognitive measures used included individual tests of embedded figures, incomplete figures, old figures, block design, memory for designs, verbal analogies, memory for objects, and vocabulary recognition. Given the large number of tests that were employed, at the preschool level, the investigators ran a factorial analysis over all the tests. The analysis revealed a general perceptual – organizational – verbal factor which served as a composite score. A main effect of treatment was obtained at 4 and 5 years on this factor score, with *atole* subjects performing significantly better than *fresco* subjects.

There was a follow-up test again when the children were 13–19 years of age. Again a battery of tests were done, including intelligence, general knowledge, literacy, numeracy, reading comprehension, vocabulary, information processing, reaction time, memory, and paired associates. Significant differences between groups were shown on four out of the ten tests, and in all cases they favored the *atole* subjects. Specifically, numeracy, general knowledge, vocabulary, and reading comprehension were higher for the **atole** subjects than those for the **fresco** subjects. In addition, intelligence scores for the fresco group were higher as socioeconomic status increased, though this effect was not seen for the atole groups.

With respect to the earlier discussion of the confounding of undernutrition and poverty in studies of children at risk, researchers are currently exploring the possible relationship between nutritional supplementation and environmental stimulation. Granthum-McGregor et al. (2014) recently reviewed studies which included a nutrition component and a child development focus, identifying only six trials of moderate-to-good quality that examined nutrition and child stimulation separately or combined. They concluded that the trials showed that the nutrition interventions usually benefited children's nutritional status and sometimes their development, while the stimulation efforts consistently benefited child development but not their nutritional status. Somewhat to their disappointment, there was little support for a synergistic interaction between supplementation and stimulation on child development outcomes, but they nevertheless proclaim an urgent need for rigorous studies which added stimulation to extant nutrition and health services.

6.7   Changes in Motor Activity

Some of the investigations mentioned earlier also measured motor abilities, and the investigators have generally found that motor skills are delayed in children with PEM (Cravioto and Arrieta 1986), although this is not always the case (Bartel et al. 1978). Regardless of the assessment of motor performance employed, undernourished infants who received nutritional interventions in the form of high-calorie feeding supplements show improved motor scores through 2 years of age. In contrast, a recent intervention in Bangladesh showed no improvement in motor scores of their severely undernourished infants, though the investigators acknowledge that their bi-weekly provision of food packets was far less of a treatment than what was routinely delivered in poor interventions.

Though PEM may be less of a threat to motor abilities than cognitive performance, it can nevertheless be of some consequence in terms of its impact on activity level. It has already been described earlier how lessened motor activity in infants may reduce their exploration of the environment and contact with stimulations. Evidence exists that school children who are only mildly undernourished can have their activity level reduced as well. In a Kenyan study, the researchers had the opportunity to observe 7-year-old children who were receiving about 95% of their recommended intake relative to their small size. Using a valid coding system, the researchers looked at playground behavior over the course of 6–8 months and separately coded low and high demonstrations of activity. They found that high activity was positively correlated with protein-calorie intake and that low activity was negatively correlated with intake (Espinosa et al. 1992). The latter association means that lower-protein-energy intake corresponded to more instances of low activity. It would seem to make sense that children who are more active require higher levels of food intake, but the data from this correlational investigation are consistent with the results of the studies that examined severely malnourished children.

6.8   Problems Connected with Behavior

Formerly malnourished children show less emotional control, are more distractable, have lower attention spans, and develop poorer relationship with their peers as well as with teachers. Families who report food insecurity – the uncertain availability of adequate and safe food – have children who are rated as higher in hyperactivity and other problematic behavior (Murphy et al. 1998). Despite cultural differences in expectations for behavior, malnourished children generally seem to have more behavioral problems than do the children with whom they are compared, even when no longer hungry.

6.9   Performance in the School

A number of studies have shown that children who were severely malnourished in infancy were found to earn poorer grades in school than matched controls. This effect is apparent even when children are matched by their class in school.

6.10   Effects in Grown-Ups

Compared to the effects of PEM in children, current knowledge on its effects in adult is rather scant. Adolescents from low-income households who experience food insufficiency report higher levels of dysthymia, or mild depression, suggesting that persistent food deprivation may affect mental health (Alaimo et al. 2002). From what is known of individuals who have experienced severe famine or survived the deprivations of a concentration camp, for example, lethargy and lessened activity are, indeed, more likely to be observed with apathy, social isolation, and impairments in memory. Binge eating has also been seen among malnourished prisoners of war (POWs) after their rescue (Polivy et al. 1994). Holocaust survivors maintain disturbed attitudes toward food (Sindler et al. 2004). Military veterans have reported hoarding and even dreaming about food after their service, especially those who were POWs, or in combat (Smith et al. 2009). Given the special circumstances of these conditions of malnutrition, however, it is impossible to determine whether such behavioral changes were directly related to PEM or to the other devastating aspects of those environments (Widdowson 1985).

The most interesting behavioral effects of malnutrition in adults comes from work conducted during the World War II. A study on the effects of moderate malnutrition to provide a basis for nutritional rehabilitation programs in Europe after the war was conducted by Keys et al. (1950). Thirty-six conscientious objectors volunteered as subjects, agreeing to participate in the experiment after being told of the purpose of the investigation. The men were housed at the University of Minnesota and were permitted to attend classes and participate in other activities. Following a baseline period during which time their daily caloric intake averaged 3200 kilocalories, the investigators reduced their energy intake by 50% to 1570 kilocalories for 6 months. While the physiological effects were of major interest to the investigators, the behavioral findings due to marked reduction in caloric intake are what made this investigation a landmark one.

The behavioral consequences of undernutrition were also assessed quantitatively through the use of questionnaires and psychological inventories and qualitatively by personal interviews and observations. One of the most important findings is that the men became introspective and found the exercise too tiring; additionally, the psychological inventories showed decreased activity, motivation, self-discipline, sex drive, and mental alertness, with enhanced apathy, irritability, and moodiness. At the completion of the investigation, nutritional rehabilitation led to reversal of the majority of behavioral changes which had been observed. Anecdotal reports suggested that even after complete rehabilitation, the subjects showed enhanced interest in food (Keys et al. 1950). More recently, anecdotal evidence from an experimental living arrangement known as Biosphere 2 supports these classic results of the above authors. Biosphere 2 was a 3 acre living space in the Arizona desert in the USA which contained an ecosystem that was materially closed but energetically open to sunlight and electricity. Air, water, and organic material were recycled in an effort to simulate what life on a space colony would be like. The four men and four women scientist/subjects who were sealed inside the station initially subsisted on 2500 calories/day. When the Biosphere was unable to produce enough food to meet the calorie requirements of the residents, the project director placed all of the residents and himself on a 1780 calorie regimen. After 6 months of calorie restriction, men lost 25 lb and women 15 lb on average, with both groups significantly reducing their cholesterol, blood pressure, and blood sugar (Walford et al. 2002). However, much like the Keys' subjects thought of the food was paramount among the subjects. The individuals reportedly stared at one another's plates at mealtime to make sure no one was getting more than his/her fair share, while any food shown in the movies at night served to mesmerize them (Schardt 2003).

References

Alaimo, K., Olson, C., & Frongillo, E. A. (2002). Food insufficiency, but not low family income, is positively associated with dysthymia and suicide symptoms in adolescents. *Journal of Nutrition, 132*, 719–725.

Barker, D. J. P. (Ed.). (1992). *Fetal and infant origins of adult disease*. London: BMJ.

Bartel, P. R., Griessel, R. D., Burnett, L. S., Freiman, I., Rosen, E. U., & Geefhuysen, J. (1978). Long-term effects of kwashiorkor on psychomotor development. *South African Medical Journal, 53*, 360–362.

Brown, J. L., & Pollitt, E. (1996). Malnutrition, poverty, and intellectual development. *Scientific American, 274*, 38–43.

Cravioto, J. (1977). Not by bread alone: Effect of early malnutrition and stimuli deprivation on mental development. In O. P. Ghai (Ed.), *Perspectives in pediatrics* (pp. 87–104). New Delhi: Interprint.

Cravioto, J., & Arrieta, R. (1986). Nutrition, mental development, and learning. In F. Faulkner & J. M. Tanner (Eds.), *Human growth* (Vol. 3, pp. 501–536). New York: Plenum Publishing.

Dover, G. J. (2009). The Barker-Hypothesis: How pediatricians will diagnose and prevent common adult-onset of diseases. *Transactions of the American Clinical and Climatological Association, 120*, 199–207.

Endres, J. B., Rockwell, R. B., & Mense, C. G. (2004). *Food, nutrition, and the young child*. New York: Macmillan Publishing Company.

Espinosa, M. P., Sigman, M. D., Neuman, C. G., Bwibo, N. O., & McDonald, M. A. (1992). Playground behaviors of school-age children in relation to nutrition, schooling, and family characteristics. *Developmental Psychology, 28*, 1188–1195.

Galler, J. R., & Ramsey, F. (1985). The influence of early malnutrition on subsequent behavioral development: The role of the micro environment of the household. *Nutrition and behavior, 2*, 161–173.

Georgieff, M. K. (2007). Nutrition and the developing brain: Nutrient priorities and measurement. *American Journal of Clinical Nutrition, 85*, 614S–620S.

Gopalan, C., & Naidu, A. N. (1972). Nutrition and fertility. *The Lancet, 300*, 1077–1079.

Gorman, K. S. (1995). Malnutrition and cognitive development: Evidence from experimental/quasi-experimental studies among the mild-to-moderately malnourished. *Journal of Nutrition, 125*, 2239S–2244S.

Granthum-McGregor, S. (1995). A review of studies of the effect of severe malnutrition on mental development. *Journal of Nutrition, 125*, 2233S–2238S.

Granthum-McGregor, S. M., Fernald, L. C. H., Kagawa, R. M. C., & Walker, S. (2014). Effects of integrated child development and nutrition interventions on child development and nutritional status. In M. M. Black & K. G. Dewey (Eds.), *Every child's potential: Integrating nutrition and early childhood development interventions* (Annals of the New York Academy of Sciences) (Vol. 1308, pp. 11–32).

Hackman, D. A., & Farah, M. J. (2009). Socioeconomic status and the developing brain. *Trends in Cognitive Sciences, 13*, 65–73.

James, W. P. T., Nelson, M., Ralph, A., & Leather, S. (1997). Socioeconomic determinants of health: The contribution of nutrition to inequalities of health. *British Medical Journal., 314*, 1545–1549.

Keys, A. J., Brozek, J., Henschel, A., Mickelson, O., & Taylor, H. L. (1950). *The biology of human starvation* (Vol. 2 vols). Minneapolis: University of Minnesota Press.

Krugman, S. D., & Dubowitz, H. (2003). Failure to thrive. *American Family Physician, 68*, 879–886.

Lester, B. M., Klein, R. E., & Martinez, S. J. (1975). The use of habituation in the study of the effects of infantile malnutrition. *Developmental Psychobiology, 8*, 541–546.

Murphy, J. M., Wehler, C. A., Pagano, M. E., Little, M., Kleinman, R. E., & Jellinek, M. S. (1998). Relationship between hunger and psychosocial functioning in low-income American children. *Journal of the American Academy of Child & Adolescent Psychiatry, 37*, 163–170.

Naismith, D. J. (1969). The foetus as parasite. *Proceedings of the Nutrition Society, 28*, 25–31.

Polivy, J., Zeitlin, S., Herman, C., & Beal, A. (1994). Food restriction and binge eating: A study of former prisoners of war. *Journal of American Psychology, 103*, 409–411.

Ricciuti, H. (1993). Nutrition and mental development. *Current Directions in Psychological Science, 2*, 43–46.

Rice, A. L., Sacco, L., Hyder, A., & Black, R. E. (2000). Malnutrition as an underlying cause of childhood deaths associated with infectious diseases in developing countries. *Bulletin of the World health Organization, 78*, 1207–1221.

Save the Children. (2012). *State of the world's mothers 2012: Nutrition in the first 1000 days.* Available at: http://www.Savethechildren.org/atf/cf%7B9def2ebe-10ae-432c-9bdO-df91d2eba74%7D/STATE-OF-THE   WORLDS-MOTHERS-REPORT-2012-FINAL.PDF. Accessed 24 Mar 2015.

Schardt, D. (2003). Eat less and live longer? Does calorie restriction work? *Nutrition Action Health Letter, 30*(1), 3–6.

Schurch, B., & Scrimshaw, N. S. (Eds.). (1990). *Activity, energy expenditure and energy requirements of infants and children.* Lausanne: Nestle Foundation.

Sindler, A. J., Wellman, N. S., & Stier, O. B. (2004). Holocaust survivors report long-term effects on attitudes toward food. *Journal of Nutrition Education and Behavior, 36*, 189–196.

Smith, C., Klosterbuer, A., & Levine, A. S. (2009). Military experience strongly influences post-service eating behavior and BMI status in American veterans. *Appetite, 52*, 280–289.

Super, C. M., Herrara, M. G., & Mora, J. O. (1990). Long-term effects of food supplementation and psychosocial intervention on the physical growth of Colombian infants at risk of malnutrition. *Child development, 61*, 29–49.

Torun, B., & Viteri, F. E. (1988). Protein-energy-malnutrition. In M. E. Shils & V. R. Young (Eds.), *Modern nutrition in health and disease* (pp. 746–773). Philadelphia: Lea & Febiger.

Townsend, J. W., Klein, R. E., Irwin, M. H., Owens, W., Yarbrough, C., & Engle, P. L. (1982). Nutrition and preschool mental development. In D. A. Wagner & H. W. Stevenson (Eds.), *Cross-cultural perspectives on child development*. San Francisco: W.H. Freeman and Co.

UNICEF. (1998). *The state of the world's children 1998: Focus on nutrition*. Available at: http://www.unicef.org/sowc98. Accessed 24 Mar 2015.

Wachs, T. D. (1995). Relation to mild-to-moderate malnutrition to human development. Correlational studies. *Journal of Nutrition, 125*, 2245S–2254S.

Wachs, T. D. (2009). Models linking nutritional deficiencies to maternal and child mental health. *American Journal of Clinical Nutrition, 89*, 935S–939S.

Walford, R. L., Mock, D., Verdery, R., & MacCallum, T. (2002). Calorie restriction in Biosphere 2: Alterations in physiologic, hematologic, hormonal, and biochemical parameters in humans restricted for a 2-year period. *Journal of Gerontology: Biological Sciences, 57*, B211–B224.

Widdowson, E. M. (1985). Responses to deficits of dietary energy. In K. Blaxter & J. C. Waterlow (Eds.), *Nutritional adaptation in man* (pp. 97–104). London: John Libby.

Winick, M., & Rosso, P. (1969). Head circumference and cellular growth of the brain in normal and marasmic children. *Journal of Pediatrics, 74*, 774–778.

Chapter 7
Mineral Deficiency and Behavior Vis-à-vis the Central Nervous System

Abstract The chapter would primarily discuss mineral deficiency and behavior *vis-à-vis* the central nervous system (CNS), and there would be discussion on the link between CNS and important trace elements like iron (Fe), zinc (Zn), and iodine (I), and additionally, there would also be a discussion on their metabolic and physiological affects

Keywords Central nervous system (CNS) · Trace elements · Behavior

The finding that minerals like zinc, iron, and copper were required in trace amounts for the normal growth and maturation of plants over a century ago led to the belief that it would be the case with humans, as well. Further evidence of their importance came to light from animal studies, where growth and reproduction were stunted when these minerals were deficient. Though over 4600 different elements are recognized as minerals by the International Mineralogical Association, only those which are more than 5 g in the human body or animal tissue are considered the major minerals (major elements) and less than 5 g are considered minor minerals (minor or trace elements). For humans, the essential major minerals in order of the amounts found in the body are calcium (Ca), phosphorus (P), potassium (K), sulfur (S), sodium (Na), chlorine (Cl), and magnesium (Mg). The body of an adult woman contains approximately 1 kg Ca, and she needs to ingest at least 1000 mg of Ca each day. By contrast the body of an adult man contains only 3–4 g of iron (Fe), making it a trace metal, and he needs about 8 mg/day. Minerals serve a variety of critical functions, as constituents of a number of enzymes and body tissues. They also act as cofactors for biological reactions and help absorption of nutrients from gastrointestinal tract and the uptake of nutrients by cells. Minerals such as Na, K, and Ca are vital for the normal functioning of the CNS. Like all living cells, nerve cells maintain an electric gradient across the membrane. At rest, the inside of the cell is slightly negatively charged relative to the outside, which is slightly positively charged – the opposite for a clay particle in the soil. This electrical gradient is regulated by the action of the Na-K pump, which controls the movement of positively charged ions of Na and K through specific channels in the membrane. The Na-K pump pushes Na into the extracellular fluid at a slightly higher rate than it permits K to cross into the cell. The net effect of this movement (as well as the contribution of other ion fluxes) is the accumulation of more positive charge on the outside of the membrane. The permeability of the membrane is momentarily altered when a nerve impulse is generated. Ion channels which are selectively permeable to Na and K, open, allowing the rapid flux of Na into the cell and K out of the cell. This leads to a temporary change in the electrical charge on the membrane (i.e., depolarization), generating an action potential. This charge then alters the permeability of the next portion of the membrane, which in turn changes the electrical charge in that portion of the membrane and so on. Thus, the nerve impulse is passed down to the fiber (Hall and Guyton 2006). When a nerve impulse reaches the synapse or end terminal of a nerve fiber, it triggers the release of a neurotransmitter, which then crosses the synaptic space to activate a second nerve fiber. We can illustrate this process using the neurotransmitter glutamate as an example. The action of the glutamate is dependent on the presence of Ca ion in the extracellular fluid. As has already been described above of the Na-K pump, specialized pumps actively remove Ca from cells so that the concentration of Ca is much higher on the outside of the cell than on the inside. Glutamate binds directly to these channels causing them to open. The momentary influx of Ca shifts the permeability of the membrane generating an action potential in the second neuron.

The above discussion proves that the neural circuitry is regulated by processes which are critically sensitive to the presence of minute concentrations of minerals in the fluids surrounding nerve cells. Factors which modify the concentrations of these ions can disrupt the integrity of neuronal membranes and interfere with the transmission of nerve impulses. Neither Na nor K is a common nutritional problem. Nevertheless, Na deficiency, known as hyponatremia, can occur in a variety of medical conditions, including chronic wasting diseases such as cancer, liver disease, semi-starvation, and ulcerative colitis, major surgical treatment, or extensive trauma as a result of abnormal external loss of Na without adequate replacement including gastrointestinal losses due to diarrhea, vomiting, and excessive sweating due to and as the result of severe dietary restriction (Heinberger 2014). On the other hand, K deficiency known as hypokalemia is often associated with abnormal intake of food which occurs as a consequence of severe malnutrition, as discussed earlier, in anorexia nervosa, chronic alcoholism, and low-carbohydrate food intake for weight loss. Hypokalemia can have profound adverse effect on neural function (Heinberger 2014).

In the case of Ca, its insufficiency will not produce CNS abnormalities, but hypocalcemia leads to hypoparathyroidism, chronic renal failure, and the intake of certain drugs like tetracycline antibiotics, which bind Ca and make it unavailable for bodily functions. Depression and dementia are other noted Ca deficiency effects.

7.1   Link Between CNS and Trace Elements

The close link between CNS and trace elements (Fe, Zn, and I) is well established. Brain development during the first 1000 days for the infant is much affected by trace elements. The following discussion pertains to this link.

7.2   Iron (Fe)

The primary function of Fe is to facilitate oxidative metabolism. Most of the body's Fe is found in the hemoglobin, the principal component of the blood cell. Hemoglobin combines with oxygen in the lungs and releases oxygen in the tissues whenever a need exists. It also aids in the return of carbon dioxide from tissues to the lungs. Fe is a structural component or a cofactor for a number of enzymes which neutralize oxidative damage of the cells, such as peroxidases, or participate in DNA and neurotransmitter synthesis and degradation.

7.3   Dietary Sources and Requirements of Iron

Hemoglobin mass, size of the organs, age, gender, and weight determine the total Fe content in the body. Average Fe content in males is 4 g, in females 2.5 g. More than 75% of Fe in the body is in the functional pool as hemoglobin mass, myoglobin, tissue enzymes, or blood (bound to the protein transferrin). The remaining amount is stored in the liver, spleen, and bone marrow as ferritin, a soluble Fe complex, or as hemosiderin, an insoluble Fe-protein complex. The normal adult male must assimilate about 1 mg/day Fe to balance the natural losses which occur via the gastrointestinal tract, urinary system, and skin. As a result of the loss of blood through menstruation, women must absorb 1.5 mg Fe/day. The important fact is that estimating the Fe adequacy of the diet by tabulating the Fe content of foods can be misleading since only a portion of the Fe present in the food can be assimilated. The bioavailability of Fe, that is, the amount that is absorbed and utilized, is influenced by a number of factors, including the composition of the diet and Fe status of the individual.

7.4   Absorption of Iron

Fe is present in food, as part of heme found in muscle fiber of meat, poultry, and fish and as non-heme found in dairy products and plant foods. The Fe from meat is well absorbed and is little affected by dietary factors. In contrast, Fe from plant foods is poorly assimilated and is strongly influenced by gastrointestinal and dietary factors. This is because ingested, non-heme Fe is insoluble. To absorb non-heme Fe, it must combine with acid gastric juice, which both liberates and stabilizes the Fe. Hence, changes in the acidity of the stomach can play an important role in the absorption of Fe. The consumption of ascorbic acid (vitamin C) in a plant-based meal can thus increase Fe absorption twofold. Conversely, decreased stomach acidity due to over-use of antacids can reduce Fe absorption.

Phytates in rice, grains, nuts, and fiber in legumes inhibit the absorption of non-heme Fe because they combine with Fe to form non-soluble compounds which pass through the intestinal tract without being absorbed. Drinking coffee or tea with a meal also decreases Fe absorption by 40% and 60%, respectively, due to the presence of Fe-biding substances called phytophenols in these beverages. Interestingly, the absorption of Fe from plant foods is more efficient when these foods are consumed with meats than when eaten alone. Cooking methods also influence Fe availability. Fe is lost from foods if they are cooked in large amounts of water, which is subsequently discarded. In contrast, cast iron cook ware can add to daily Fe intake. It must be understood that it is the Fe status of an individual which alters Fe absorption. Fe-deficient individuals are twice as efficient in absorbing Fe than non-deficient individuals. Although heme Fe is approximately 2–3 times better absorbed than non-heme Fe, the majority of Fe consumed in typical mixed diets in the USA and Canada is predominantly in the non-heme form. Therefore, for the purpose of setting the dietary requirement for Fe, the bioavailability of ingested Fe from mixed diets is estimated to be 18%. The bioavailability of Fe from a vegetarian diet is approximately half that of a mixed diet (approximately 10%). Thus, vegetarians may have difficulty consuming adequate Fe (Food and Nutrition Board 2001a).

7.5   The Dietary Sources for Iron

Though liver (from sheep and beef) is a good source of Fe, it has fallen into disfavor, of late, because of its high cholesterol content. Lean meats, shellfish, and poultry are other good sources. Eggs, leafy vegetables, whole-grain cereals, and fruits provide progressively less absorbable Fe (Food and Nutrition Board 2001a). Current dietary patterns may have a negative impact on Fe status. Approximately half the Fe in the average American diet is from bread or grain products from which Fe is poorly absorbed. The consumption of pizza, snack foods, and soft drinks is increasing, but these foods provide only negligible amounts of usable Fe. Fortification of bread with Fe has been done in western diets, and Venezuela had success in reducing Fe deficiency anemia through fortification by half in less than 2 years. A recent innovation is the use of "sprinkles," single-dose Fe sachets (ferrous fumarate) or additional micronutrients in a powdered form that can be sprinkled on any food item at home. As compared to the traditional use of Fe drops, sprinkles showed greater ease of use

Table 7.1 The Fe content of some foods/serving

Food	Serving	Fe (mg)
Enriched cornflakes	One cup	8.4
Beef liver	Three ounces	5.2
Green peas	One cup	3.2
Raisins	One cup	2.7
Ground beef	Three ounces	2.2
White rice	One cup	1.9
Tuna fish	Three ounces	1.3
Green beans	One cup	1.2
Chicken	Three ounces	1.0
Whole wheat bread	One slice	0.9
Fried egg	One large	0.9
Baked potato	One medium	0.6
Tomatoes	One cup	0.5
Cheddar cheese	One ounce	0.2

Source: USDA (United States Department of Agriculture)

and adherence, less staining of teeth, and, most importantly, a higher reduction of anemia in infants and toddlers (Christofides et al. 2006) (Table 7.1).

7.6 Fe Deficiency

Fe deficiency can be categorized as follows:

Involves depletion of Fe stores and is evidenced by a decrease in the concentration of serum ferritin.

There is reduction in body transport and an increase in absorption. Although hemoglobin production is compromised in Fe deficiency, no functional impairments are observed.

Development of anemia, leading to weakness, fatigue, breathing difficulty during exercise, headache, and palpitations.

7.7 Fe Deficiency and Behavior

In children, Fe deficiency anemia is strongly associated with impaired cognitive development and intellectual impairment. The behavioral aberrations in both children and infants include irritability, mental fatigue, shortened attention span, impaired memory, anxiety, and depression.

It was during the 1980s and 1990s that much valuable information was gathered on the effects of Fe deficiency in infants and adults in Central America, Africa, and elsewhere around the globe. Investigations in infants have used the Bayley Scales of Infant Development (please refer to earlier chapter) which measures mental and motor abilities during the first 2 years of the infant's life, as well as temperament. Fe-deficient infants generally score lower on the Bayley Scales than non-deficient infants. The developmental test scores were significantly lower in anemic children 6–24 months old, compared to the others who are not, in Costa Rica (Lozoff et al. 1998). Additionally, anemic children were found to be more fearful, hesitant, and less active and show less pleasure than non-anemic children. As with PEM, Fe deficiency often occurs in the context of poverty and among families coping with a variety of stress factors, like large family size, food insecurity, poor housing, low maternal education, and poor mental health. Thus, a variety of socioeconomic factors can confound the association between Fe deficiency and delayed infant development.

A crucial question in behavioral science is whether Fe deficiency in early life results in long-standing cognitive and behavioral deficits in children. The relationship between Fe deficiency and cognitive test scores were measured in approximately 5000 children, a nationally representative sample, and it was found that average math scores were lower for children with Fe deficiency (either with or without anemia) than in children with normal Fe status (Halterman et al. 2001). Moreover, Fe-deficient children were more than twice as likely to score below average score in math compared to children with adequate Fe status. More strikingly, with investigations as follow-up work with their Costa Rican cohort children who had been tested and treated for Fe deficiency in infancy, it was observed that at 12 years of age, children who were Fe-deficient as infants lagged behind their peers with good Fe status, in math, writing, reading, and overall school progress (Lozoff et al. 1998). Children who had been Fe-deficient in infancy also experienced more anxiety, depression, and social problems, as their performance was inferior in cognitive and motor tasks (Lozoff et al. 2000). At age 19, the Fe-deficient children could not catch up with children who had good Fe status, widening the gap for those of lower socioeconomic status. These results clearly establish the distinctive role of Fe in behavior. Regarding the effect of Fe therapy, it was seen that the effect of Fe deficiency can be reversed, by a 4 month Fe therapy, as seen in the case of toddlers through the Bayley mental psychomotor scales, in Indonesia (Idjradinata and Pollitt 1993). In a recent review of studies that examined the benefits and risks for Fe supplementation, in children under 6 years of age, it was observed that Fe supplements from 2 to 6 months duration were sufficient to improve performance by infants on the Bayley scales, though such global tests may miss subtle differences in neurobehavioral development. In conclusion, it seems that Fe deficiency is associated with a range of deficits in brain functioning and cognitive behavior. This represents a major public health concern, not only in underdeveloped countries of Africa and Asia but also in countries like the USA, where the prevalence of Fe deficiency ranges from 6% to 14% among toddlers aged 1–3 years (Baker et al. 2010).

7.8   Zinc (Zn)

Zn is the most important micronutrient whose deficiency is being manifested globally, in both human and animal nutrition. It has emerged as the most important micronutrient limiting crop yields, and its importance and factors which determine its soil availability have been discussed in great detail (Nair 2016). Zn is a cofactor for nearly 100 enzymes which catalyze vital biological functions. It facilitates the synthesis of DNA and RNA and thus participates in protein metabolism. Zn is, in particular, important for protein metabolism in tissues which undergo rapid turnover, such as gonads, gastrointestinal tract, taste buds, and skin. It also plays a critical function in immune system, physical growth, and the actions of a number of hormones. As Zn plays an essential role in ever so many diverse biologic processes, deficiencies of the nutrient lead to a broad range of non-specific deficits. Hence, it will be difficult to link a specific clinical abnormality directly to Zn by itself (Food and Nutrition Board 2001b).

Though Zn is widely distributed in foods, its bioavailability, like that of non-heme Fe, depends upon several gastrointestinal factors, including intestinal pH and the presence or absence of other food components, such as phytates (as in the case of Fe), which interfere with Zn absorption. The Zn absorption cycle into the human body has a two-step mechanism: first, its bioavailability in the soil, consequently to the plant root for subsequent absorption, and then its bioavailability within the body after absorption. That Zn availability in the soil is exclusively a function of soil pH, as has been thought for a long time, has been disproved by the latest research of Nair (2016), where he showed that it is the actual Zn "buffer power" which decides its bioavailability in soil and consequently to the plant (Nair 2013a, b). There are thermodynamic reactions involved in this, absorptive process. Nair (1996) has experimentally proved this. Zn also competes with other elements like Fe and Ca for absorption. High intake of Fe as dietary supplements can inhibit Zn absorption (Food and Nutrition Board 2001b).

As seen in the data in Table 7.2, good sources of Zn include seafood (oysters), muscle, meat, and nuts. Legumes and whole-grain products contain significant amounts of phytates which decrease the amount of available Zn in these foods. Fruits and vegetables are low in Zn. The relatively minimal amounts of absorbable Zn in plant foods make them an inadequate source of Zn for vegetarians. Zn requirements may be as much as 50% higher for strict vegetarians because of the relatively poor absorption of Zn from plant foods (Food and Nutrition Board 2001b).

No adverse effects have been noted because of normal Zn uptake. However, high uptake of Zn from Zn supplements can cause gastrointestinal distress and can lead to interference with copper status. For adults, 40 mg/day from food, water, and Zn supplements is adequate (Food and Nutrition Board 2001b).

Table 7.2 The Zn content of some foods/serving

Food	Serving	Zn (mg)
Oysters	4 ounce	74.1
Ground beef patty	3 ounce	5.3
Cooked ham	3 ounce	2.2
Canned kidney beans	One cup	1.4
Frozen green peas	One cup	1.1
Canned corn	One cup	1.0
Roasted chicken breast	3 ounce	0.9
Peanut butter	Two table spoon	0.9
Cheddar cheese	One ounce	0.9
Cooked broccoli	One cup	0.7
Whole wheat bread	One slice	0.5
Baked salmon	Three ounce	0.4
Canned tuna fish	Three ounce	0.4
Apple	One, medium sized	0.1

Source: USDA (United States Department of Agriculture)

7.9   Zinc Deficiency: Human Investigations

Zinc deficiency is now global. It was in the early 1970s when the deficiency symptoms first showed up in crops like maize and wheat. In Punjab State, in India, the "cradle" of the so-called green revolution, despite huge quantities of nitrogen, phosphorus, and potassium application, the plants were still stunted. Fifteen kilograms of Zinc sulfate did the miracle, as the soil was depleted of its native zinc due to the continuous cultivation of the hybrid rice and wheat. More recently, this author observed extreme cases of zinc deficiency in Central Asia (Turkey). Farmers were applying as much as 100 kg zinc sulfate per hectare with no effect until a systematic research project set up by this author showed that it is the "Zinc Buffer Power" of different soils that affected zinc availability (Nair 2016). These results are discussed in detail, in a chapter, in the prestigious publication *Advances in Agronomy* (Nair 2016).

Symptoms of Zn deficiency were first noticed in the 1960s in the Middle East, and its deficiency in humans is prevalent throughout the developing world. Approximately more than one-third of the world's population suffers from suboptimal Zn status. In the USA severe Zn deficiency is rare, but mild deficiency has been associated with strict vegetarians.

7.10   Acrodermatitis Enteropathica (AE)

The disease is a rare genetic disorder which is inherited as an autosomal recessive characteristic. Individuals with AE are unable to absorb Zn from the gastrointestinal tract leading to functional Zn deficiency (King and Cousins 2014). The underlying biochemical defect responsible for AE is unknown. The onset of the disease occurs in infants at weaning time and is characterized by acute growth retardation; severe gastrointestinal problems, including diarrhea and malabsorption; hyper-pigmented skin lesions; and increased susceptibility to infection. Alterations in mood and behavior are also a common feature of AE. Children with the disease are generally lethargic and irritable and rarely smile or display an interest in their environment. They may also suffer from emotional disorders including depression. If undiagnosed and untreated, AE can often turn fatal, with resulting death from infection and/or malnutrition. Zn supplements rapidly reverse the symptoms (King and Cousins 2014).

7.11 Nutritional Zn Deficiency

The earliest zinc deficiency was detected in soils and plants in the1950s and 1960s (Nair 1996), and the symptoms in humans came much later. The first instance of severe Zn deficiency was described in young Egyptian and Iranian men suffering from growth retardation (Prasad 1988). Severe Zn deficiency in West Asian soils (Turkey) was reported by Nair (2013a, b). In the case of the report by Prasad (1988), many of these men looked like young adolescent boys, although they were in their early twenties. The report of Nair (2013a, b) showed how his revolutionary soil management concept, now globally known as "The Nutrient Buffer Power Concept," could tackle the problem at the soil level, because conventional and textbook knowledge of soil management had failed to correct the severe deficiency in wheat plants. The affected men, in addition to retarded growth, also displayed retarded sexual maturation, anemia, dermatitis, impaired liver function, anorexia, and neurosensory and behavioral abnormalities. The majority of affected people came from the lowest socioeconomic strata of their villages and lived on cereal-based diets consisting mainly of unleavened bread and vegetables. Meat and dairy products were rarely available for these deprived populations. Parasitic infection, which can interfere with Zn absorption and are common among these populations, might have contributed to the deficiency. The effects of Zn supplements on these men were striking. Abnormalities in sensory function are also common among Zn-deficient individuals, including a decreased sense of taste (hypogeusia), night blindness, and photophobia. One report showed that prescription of Zn supplements to such people cured the symptoms in 6 months (Tanaka 2002). The mechanism by which Zn deficiency affects taste function is poorly understood but may be related to pathological changes in taste bud anatomy via disruptions in salivary protein known as gustin (Henkin et al. 1999). Gustin is a Zn-binding enzyme which plays a crucial role in the growth and development of taste buds. Factors which reduce gustin secretion, such as Zn deficiency, or the functionality of the gustin protein (such as a recently described genetic mutation which reduces the ability of gustin to bind Zn), adversely affect the health of taste buds (Melis et al. 2013).

7.12   Effect of Zinc on Brain Development and Behavior

Several investigations have focused on this aspect. Working in a semi-rural Egyptian population, Kirksey et al. reported a positive association between maternal intake of foods high in Zn content and infant performance on attention tasks soon after birth. At 6 months of age, motor development was also associated with maternal intake of Zn during gestation (Kirksey et al. 1994). The effects of Zn supplementation on brain development and behavior are somewhat controversial. Supplementation trials have generally produced mixed results with some studies showing improvements in cognitive development and learning skills while others showed no benefit (Gibson et al. 1989). Since Zn deficiency can coincide with deficiencies in other trace elements, it is often difficult to isolate the specific effects of Zn on behavioral outcome. Sanstead et al. (1998) investigated the effects of Zn therapy on cognitive performance and psychomotor function in a large Chinese population of children aged 6–9 years, half of which is from urban area and the other half from rural area. Subjects participated in a 10-week supplementation trial and received either Zn alone, a mineral supplement (without Zn), or a Zn-mineral supplement. Results showed that Zn supplementation either with or without other minerals resulted in improved performance in a complex reasoning task. Other neurophysiological functions were also improved, including recognition memory, attention, and psychomotor function.

7.13   Iodine (I)

The importance of iodine (I) in both plant and human nutrition was noted much later than in the case of trace elements like zinc. In humans, iodine is an indispensable component of the thyroid hormones. Thyroid hormones regulate the rate of cellular oxidation, thereby influencing physiological and mental development, the functioning of nervous and muscle tissues, and energy metabolism. It is important for normal maturity of the CNS and fetal development. Goiter or enlargement of the thyroid gland is the hallmark of I deficiency (Lauberg 2014).

7.14   I Requirements and Dietary Sources

Nearly 75% of the 15–25 mg of I in the body is concentrated in the thyroid gland with the remainder found in the salivary and mammary glands, gastric mucosa, and kidneys. I is found in both plant and animal foods. However, the I content of foods depends primarily on the I content of the soil on which the plants, from which the food is derived, are grown, and, therefore, is highly variable. Soils from areas near oceans, which have a high concentration of I, typically contain large amounts of

I. In contrast, mountainous parts of the world, such as the Andes, Alps, Pyrenees, and Himalayas, have soils from which most of the I has been removed through the natural process of glaciation, weathering, and erosion. I-deficient soils are also found in flooded river valleys such as the Ganges, in India. In the USA, I-deficient soils are found in areas surrounding the Great Lakes. This region was considered the "Goiter Belt" prior to the iodination of the salt (Lauberg 2014).

Inasmuch as food sources for I are considered, with the exception of shellfish, saltwater fish, and seaweed, which are high in I content, most other food sources are poor sources of I. Iodized salt is the source of dietary I for many countries. In the USA, iodized salt contains 0.01% potassium iodide or 76 microgram iodine/g (Food and Nutrition Board 2001c).

7.15 Iodine Deficiency, Metabolism, and Physiology

There is a rapid circulation of I subsequent to absorption from the gut. Most of absorbed I is used by the thyroid gland to make thyroid hormones. Remainder is absorbed by the kidney and excreted in urine. The thyroid gland is extremely efficient in extracting I from blood and concentrating it, and the process is called "iodine trapping." The thyroid gland is regulated by a complex mechanism involving the thyroid hormones, the pituitary gland, the brain, and peripheral tissues. When blood levels of thyroid hormones fall, the pituitary gland is stimulated to release thyroid-stimulating hormone TSH. TSH then travels to the thyroid gland to increase the rate of I uptake and hormone synthesis. When I status is adequate, hormone synthesis increases and TSH levels fall, restoring normal hormonal balance. In I deficiency, TSH levels remain elevated. The thyroid gland responds by increasing I turnover, but it cannot keep pace with the demands for increased hormone synthesis. Cellular hyperplasia follows leading to an enlargement of the thyroid gland, known as goiter. The thyroid gland in a normal individual weighs about 15 g (Lauberg 2014). Goiter can be detected by physical examination as a mass or swelling in the neck and in severe cases may be clearly visible as in the following photo (Fig. 7.1).

I deficiency can defined as mild, moderate, or severe. Some of the symptoms associated with I deficiency are shown in Table 7.3.

Moderate I deficiency is associated with reduced visual and motor performances, perceptual abnormalities, and reduced intellectual capabilities in both children and adults (Table 7.3). Severe I deficiency causes goiter in adults and children. It causes cretinism which is the result of severe I deficiency during fetal development and is associated with permanent physical and mental abnormalities. There are two types of cretinism, neurological and myxedematous. Individuals with neurological cretinism, which is more common, typically are deaf-mute and display spastic movement and gait, as well as profound mental retardation. In comparison, myxedematous cretins are short in stature and display symptoms of severe hypothyroidism, but they are not deaf or mute. The two types of cretinism tend to predominate in different

Figure 7.1 African
woman with goiter. (Photo
credit: Collection/Getty
Images)

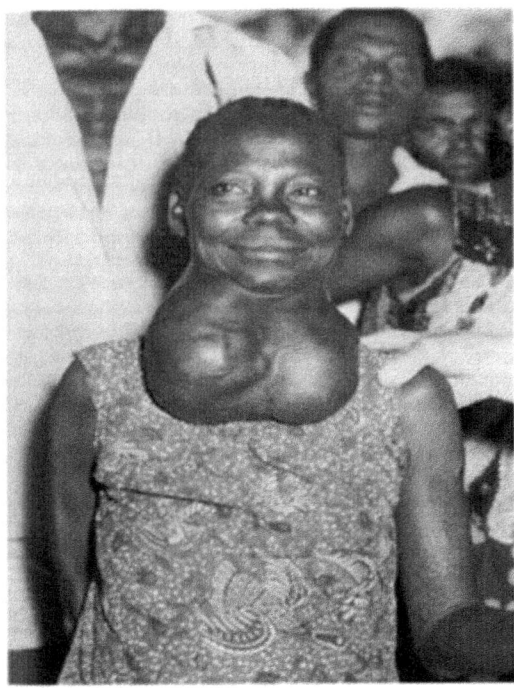

Table 7.3 I deficiency symptoms by age group

Age group	Symptoms
Fetus	Spontaneous abortion
	Still birth, congenital anomalies
	Perinatal mortality
Neonate	Infant mortality, endemic cretinism
Child and adolescent	Goiter, impaired mental function
	Delayed physical development
Adult	Goiter, impaired mental function
	Reduced work productivity, hypothyroidism

Source: World Health Organization

geographical areas, but they are not mutually exclusive. Mixed forms of cretinism with both neurological and myxedematous features do occur (Lauberg 2014).

While inadequate intake of I is the primary cause of goiter, certain foods contain substances called goitrogens which interfere with the production of thyroid hormones. Many of these foods, such as cassava, maize, bamboo shoots, millets, and sweet potato – the dietary staples of the African continent – lead to I deficiency.

Only recently the complex etiology of endemic goiter and cretinism has been unraveled, and the probable involvement of selenium deficiency and goitrogen overload. Endemic cretinism occurs when I intake is below 2.5 micrograms/day. In severely affected areas of India, Indonesia, China, and elsewhere, endemic cretinism can affect up to 10% of the population (Lauberg 2014). Moreover, in severely affected communities, a substantial proportion of the population may also have lesser degrees of retardation as a result of brief or intermittent I deficiency during early development. These individuals are not so readily identifiable because they lack the physical characteristics and developmental impairments of cretinism. For example, in rural Ecuador in addition to the 5.7% of the adult population who were deaf-mute cretins, another 17.4% of the population displayed more moderate neurological deficits and behavioral limitations (Greene 1994). Thus, the presence of large subgroups of people exhibiting a spectrum of physical and/ or intellectual limitations can adversely affect the social and economic well-being of a community.

Supplementation of I can completely eliminate I deficiency. Intake of iodized salt is the preferred route to correct I deficiency, although iodized salt injections have been successfully used in some high-risk populations. Due to the combined efforts of numerous governments and international agencies, substantial progress has been made in implementing a universal salt iodization policy to eliminate I deficiency as a public health problem worldwide. However, significant challenges remain to ensure adequate supply and distribution of iodized salt globally, as it is estimated that I deficiency still affects one-third of the worldwide population (Lauberg 2014).

It was originally believed that thyroid hormone was not important for early brain development. It is now known that the fetal brain needs thyroid hormone throughout gestation for normal development (Delange 2000).

Numerous investigations have shown the relationship between early I deficiency and intellectual development in children. In an early study, Bautista et al. (1982) found no difference between supplemented and non-supplemented 5–12-year-old Bolivian schoolchildren on the Stanford-Binet intelligence test. However, Bleichrodt and Born (1994) conducted a meta-analysis of 19 studies on I deficiency and cognitive function and showed that I deficiency resulted in a striking loss of 13.5 points in intelligence quotient (IQ) scores. Tiwari et al. (1996) found that preadolescent and adolescent boys in Padrauna region in India with severe I deficiency were slow learners and showed low motivation to achieve. An I supplementation study in the Republic of Benin (formerly Dahomey) with 9–10-year-old schoolchildren showed improvements with treatment on a number of nonverbal measures of intelligence, including the Raven Progressive Matrices. Quite clearly, I deficiency can lead to a marked loss of intellectual potential in children.

References

Baker, R. D., Greer, F. R., & the Committee on Nutrtion. (2010). Clinical report – Diagnosis and prevention of iron deficiency and iron-deficiency anemia in infants and young children (0–3 years of age). *Pediatrics, 126,* 1040–1050.

Bautista, A., Barker, P. A., Dunn, J. T., Sanchez, M., & Kaiser, D. L. (1982). The effects of oral iodized oil on intelligence, thyroid status, and somatic growth in school-age children from an area of endemic goiter. *American Journal of Clinical Nutrition, 35,* 127–134.

Bleichrodt, N., & Born, M. (1994). A meta-analysis of research on iodine and its relationship to cognitive development. In J. Stanbury (Ed.), *The damaged brain of iodine deficiency.* New York: Cognizant Communication.

Christofides, A., Asante, K. P., Schauer, C., Sharief, W., Owusu-Agyei, S., & Zlotkin, S. (2006). Multi-micronutrient Sprinkles including a low dose of iron provided as microencapsulated ferrousfumarate improves haematologic indices in anaemic children: A randomized clinical trial. *Maternal and Child Nutrition, 2,* 169–180.

Delange, F. (2000). The role of iodine in brain development. *Proceedings of the Nutrition Society, 59,* 75–79.

Food and Nutrition Board, Institute of Medicine, National Academy of Sciences. (2001a). Iron. In *Dietary reference intakes for vitamin A, vitamin K, arsenic, boron, chromium, copper, iodine, iron, manganese, molybdenum, nickel, silicon, vanadium and zinc* (pp. 290–393). Washington, DC: National Academy Press.

Food and Nutrition Board, Institute of Medicine, National Academy of Sciences. (2001b). Zinc. In *Dietary reference intakes for vitamin A, vitamin K, arsenic, boron, chromium, copper, iodine, iron, manganese, molybdenum, nickel, silicon, vanadium, and zinc* (pp. 442–501). Washington, DC: National Academy Press.

Food and Nutrition Board, Institute of Medicine, National Academy of Sciences. (2001c). Iodine. In *Dietary reference intakes for vitamin A, vitamin K, arsenic, boron, chromium, copper, iodine, iron, manganese, molybdenum, nickel, silicon, vanadium, and zinc* (pp. 258–289). Washington, DC: National Academy Press.

Gibson, R. S., Vanderkooy, P. D. S., MacDonald, A. C., Goldman, A., Ryan, B. A., & Berry, M. (1989). Growth limiting mild zinc-deficiency syndrome in some southern Ontario boys with low height percentiles. *American Journal of Clinical Nutrition, 49,* 1266–1273.

Greene, L. (1994). A retrospective view of iodine deficiency, brain development and behavior from studies in Equador. In J. Stanbury (Ed.), *The damaged brain of iodine deficiency.* New York: Cognizant Communication.

Hall, J. E., & Guyton, A. C. (2006). *Textbook of medical physiology.* St Louis: Elsevier Saunders.

Halterman, J., Kaczorowski, J., Aligne, C., Auinger, P., & Szilagyi, P. (2001). Iron deficiency and cognitive achievement among school-aged children and adolescents in the United States. *Pediatrics, 107,* 1381–1386.

Heinberger, D. C. (2014). Clinical manifestations of nutrient deficiencies and toxicities. In A. C. Ross, B. Caballero, R. J. Cousins, K. L. Tucker, & T. R. Ziegler (Eds.), *Modern nutrition in health and disease* (11th ed., pp. 757–770). Philadelphia: Lippincott Williams and Wilkins.

Henkin, R., Martin, B., & Agarwal, R. (1999). Decreased parotid saliva gustin/carbonic anhydrase VI secretion: An enzyme disorder manifested by gustatory and olfactory dysfunction. *American Journal of Medical Sciences, 318,* 380–391.

Idjradinata, P., & Pollitt, E. (1993). Reversal of developmental delays in iron-deficient anemic infants treated with iron. *The Lancet, 34,* 1–4.

King, K., & Cousins, R. J. (2014). Zinc. In A. C. Ross, B. Caballero, R. J. Cousins, K. L. Tucker, & T. R. Ziegler (Eds.), *Modern nutrition in health and disease* (11th ed., pp. 189–205). Philadelphia: Lippincott Williams and Wilkins.

Kirksey, A., Wachs, T., Yunis, F., Srinath, U., Rahmanifar, A., McCabe, G., Galal, O., Harrison, G., & Jerome, N. (1994). Relation of maternal zinc nutriture to pregnancy outcome and infant development in an Egyptian village. *American Journal of Clinical Nutrition, 60,* 782–792.

Lauberg, P. (2014). Iodine. In A. C. Ross, B. Caballero, R. J. Cousins, K. L. Tucker, & T. R. Ziegler (Eds.), *Modern nutrition in health and disease* (11th ed., pp. 217–224). Philadelphia: Lippincott Williams and Wilkins.

Lozoff, B., Klein, N., Nelson, E., McClish, D., Manuel, M., & Chacon, M. (1998). Behavior of infants with iron deficiency anemia. *Child Development, 69*, 24–36.

Lozoff, B., Jimenez, E., Hagen, J., Mollen, E., & Wolf, A. (2000). Poorer behavioral and developmental outcome more than 10 years after treatment for iron deficiency in infancy. *Pediatrics, 105*, e51.

Melis, M., Atzori, E., Cabras, S., Zonza, A., & Calto, M. C. (2013). A gustin gene polymorphism as a mechanistic link between PROP tasting and taste papilla density and morphology. *PLoS One, 8*, e74151.

Nair, K. P. P. (1996). The buffering power of plant nutrients and effects on availability. *Advances in Agronomy, 57*, 237–287.

Nair, K. P. P. (2013a). *The agronomy and economy of turmeric and ginger*. London/Waltham: Elsevier.

Nair, K. P. P. (2013b). The buffer power concept and its relevance in African and Asian soils. *Advances in Agronomy, 121*, 447–516.

Nair, K. P. P. (2016). *The nutrient buffer power concept for sustainable agriculture* (p. 434). Chennai: Notion Press.

Prasad, A. (1988). Clinical spectrum and diagnostic aspects of human zinc deficiency. In *Essential and toxic trace elements in human health and disease*. New York: Alan R. Liss.

Sandstead, H., Penland, J., Alcock, N., Dayal, H., Chen, X., Li, J., Zhao, F., & Yang, J. (1998). Effects of repletion with zinc and other micronutrients on neuropsychologic performance and growth of Chinese children. *American Journal of Clinical Nutrition, 68*, 470S–475S.

Tanaka, M. (2002). Secretory function of the salivary gland in patients with taste disorders or xerostomia: Correlation with zinc deficiency. *Acta-Oto-laryngology, 122*, 134–141.

Tiwari, B. D., Godbole, M. M., Chattopadhyay, N., Mandal, A., & Mithal, A. (1996). Learning disabilities and poor motivation to achieve due to prolonged iodine deficiency. *American Journal of Clinical Nutrition, 63*, 782–786.

Chapter 8
Role of Dietary Supplements on Mental Function

Abstract The chapter would principally discuss the role of dietary supplements on mental function. There would also be a discussion on the definition of dietary supplements, the norms of US Food and Drug Administration 1994, safety of dietary supplements, the role of fish oil supplements, and the interrelationship between fish oil supplements and cognitive decline. The place of herbals and botanicals in human health will also form part of the discussion in the chapter. Additionally, there would also be a discussion on the various antioxidants.

Keywords Dietary supplements · Mental function · US Food and Drug Administration · Safety of dietary supplements · Science and hype

There is now a growing interest, among the public, in alternative medicine. Obviously, this is tapped by commercial interests, for profit motive. It is the consumer interest in dietary supplements which is at the center of the popularity of self-help and alternative medicines. Until about the mid-1990s, the most important dietary supplements purchased from drug stores were multivitamin tablets, even now they constitute dietary supplements, in addition to medicinal herbs, fish oils, and omega-3 fatty acid supplements. While this became the norm in the west, like the USA, since the last about two decades, countries like India are catching up fast. Even within India, there are differences among the states. For example, the highly literate Kerala State, in south India, has caught on very fast, while a state like Bihar, in eastern India, is lagging far behind. Literacy has to do much with this development. In fact, sales of vitamin/mineral formulations now account only about one-fifth of all supplements, while the total sales has skyrocketed to more than US$11.5 billion during a single generation in the USA. Sales of herbal formulations alone totaled more than US$594 million in 2012, with five top-selling herbs in the USA currently listed as cranberry, garlic, saw palmetto, soy, and gingko. But, more surprisingly, sales of fish oils and omega-3 fatty acid supplements – a market virtually nonexistent some three decades ago – are now at US$ 1043 million/year, nearly twice that of herbals (Grey and Bolland 2014).

Concomitant to this development, or perhaps preceding it, is a new awareness about changes in life style, yoga, meditation, and other mentally stimulating activi-

© Springer Nature Switzerland AG 2020   99
K. P. Nair, *Food and Human Responses*,
https://doi.org/10.1007/978-3-030-35437-4_8

ties that endow much good to a healthy life. Perhaps, the increasing switch to the age old Indian medicine, *Ayurveda* (a word in Sanskrit, the ancient Indian language, which means the gospel of good health or the last word in good health), away from allopathic and invasive medicine, coupled with an increasing trend in vegetarianism, is a new wave that will only grow with the passage of time. The spread of spirituality is piloting this movement. Everyone wants to lead a healthy and long life, away from binge eating and drinking, a definite "no" to gluttony, seems to be the new emerging order of the day in the west. And, more people in the west are looking to India for inspiration in different walks of life.

The question whether the claims of many dietary supplements, such as enhancing mood and memory to decreasing anxiety and stress and improving sleep, are fantasy or fact merits examination. Do the benefits outweigh the risks involved? Or, are the consumers eager to embrace unproven therapies? Some of the primary reasons for using dietary supplements are to preserve overall health and well-being and to reduce the adverse effects of aging. Some supplements show promise in slowing cognitive decline in the elderly and reducing the severity of neurodegenerative diseases, such as the Alzheimer's and Parkinson's diseases. On the other hand, the extent to which dietary supplements can improve cognitive function in otherwise healthy people is open to debate. Consumers may also use supplements as an alternative to mainstream medicine or because they have exhausted conventional treatment options. For example, St. John's wort has been promoted as beneficial in the treatment of mild depression with fewer side effects than with conventional pharmaceutical therapies. The public perceives these products as "natural" and "healthy" and, therefore, without health risks. However, some botanicals can have side effects or can interfere with the actions of other medicines. The long-term effects of these products are, however, unknown. Moreover, dietary supplements are sold as foods, not as drugs. According to current legal statutes, dietary supplements do not have to meet the same rigorous safety standards applied to the manufacture and sale of established allopathic drugs. The US Dietary Supplements Health and Education Act of 1994 opened the door for supplements to be sold without safety testing, meaning that such supplements are deemed safe until someone gets sick due to their use. Yet, serious health consequences or death have resulted from the use of certain supplements (Cohen 2012). For instance, vitamins E and C, β-carotene, and selenium are consumed as dietary supplements. However, as opposed to herbs and botanicals, the functions of these nutrients in the human body are better understood, and safety levels have been established. Whether ingesting these constituents in large quantities is effective or not is an ongoing research question. This is particularly important in light of some evidence that vitamins consumed as supplements may not provide the same health benefits as consuming the same vitamins from foods. There are hundreds of dietary supplements sold in the USA and Western Europe.

8.1   Definition of a Dietary Supplement: Norms of US Food and Drug Administration 1994

According to the US Dietary Supplement Health and Education Act (DSHEA) of 1994, the term "dietary supplement" denotes the following:

A product (other than tobacco) which is intended to supplement the diet which bears or contains one or more of the following ingredients:

A vitamin
A mineral
A herb or other botanicals
An amino acid
A dietary substance for use by humans to supplement the diet by increasing total dietary intake of enzymes, or tissues, from organs or glands or a concentrate, a metabolite, constituent, or extract

It is intended for ingestion in pill, capsule, softgel, gelcap, tablet, liquid, or in powder form.

It is not represented for use as a conventional food or as the sole item of a meal or diet.

It is labeled as a "dietary supplement."

8.2   Safety of Dietary Supplements

The Food and Drug Administration (FDA) of the USA is the final authority to vouch for the safety of drugs, but dietary supplements need not be declared as safe prior to marketing. If the dietary supplement is suspected to contain a harmful ingredient, the onus of proving it rests with the FDA. In other words, the FDA cannot remove a product from the market unless there is compelling evidence that the supplement can cause harm to human life. This is in stark contrast to pharmaceuticals which are highly regulated by FDA. Currently there are no regulations that establish a minimum standard of practice for manufacturing dietary supplements to ensure their safety, purity, identity, quality strength, and composition. However, in the USA, the nonprofit group US Pharmacopeia (USP) has developed a voluntary national certification program for dietary supplements. This program will verify that a supplement contains the declared ingredients on the product label at the amounts specified, meets requirements for limits on contaminants, and complies with good manufacturing practices.

8.3   Science and Hype: The Urgent Need to Separate the Chaff from the Grain – One from the Other

There is a plethora of information on dietary supplements, which is directed at the unsuspecting public, and hence, there is a need to separate science from hype. Hundreds of clinical trials have been conducted on the efficacy of these substances, but bulk of this research is flawed because of the inappropriate methodology employed. A trial on *Ginkgo biloba* for the treatment of Alzheimer's disease without randomization of the treatments is a notable example of poor and inappropriate methodology.

8.4   What Can Dietary Antioxidants Achieve in Combating Oxidative Stress?

Aerobic metabolism produces energy, waste products such as carbon dioxide, and a steady stream of reactive products. All cells of the body are exposed to oxidants generated by normal metabolic processes and from a variety of environmental toxins, such as plant alkaloids, pesticides, air pollutants, and heavy metals. The principal reactive species include reactive oxygen species (ROS) and reactive nitrogen species (RNS). Approximately 1–3% of the total oxygen utilized by the body is involved in the production of ROS, which includes oxygen radicals, such as superoxide, hydroxyl, and peroxyl, and non-radicals such as hydrogen peroxide. RNS includes nitric oxide and several related compounds. ROS and RNS, also called "free radicals," have a free and unpaired electron which makes them highly unstable and reactive with biologic molecules, such as proteins, lipids, carbohydrates, and nucleic acids. As described below, this reactivity can damage cells and may contribute to the aging process and the development of chronic diseases.

The body has two antioxidant defense mechanisms which counteract the damaging effects of reactive species. The first mechanism involves free radical scavenging to remove ROS, RNS, and their precursors before they damage cells and cellular components. The antioxidant vitamins, such as vitamin C (ascorbic acid), vitamin E (α-tocopherol), and β-carotene, function as free radical scavengers in the human body. The second mechanism involves enzymatic processes which neutralize ROS and RNS to harmless compounds. The major protective enzyme systems include superoxide dismutases, glutathione peroxidases, and catalases. Selenium is a weak antioxidant on its own but is an essential component of the detoxifying enzyme glutathione peroxidase (Food and Nutrition Board 2000). An imbalance between free radicals (prooxidants) and antioxidants causes oxidative stress. An imbalance favoring prooxidants leads to oxidative damage to a variety of biologic molecules including lipid peroxidation of cell membranes, DNA strand breakage, and inactivation of cellular proteins. It has been hypothesized that over time, oxidative stress may play a role in the aging process because this imbalance produces damage at a faster rate than the body's repair processes can cope with. Oxidative stress has also been implicated in the development of degenerative diseases such as cardiovascular disease, stroke, and senile dementia. Although there is little doubt that oxidative stress directly damages biologic molecules, it has not been firmly established that degenerative diseases result from imbalance between formation and removal of reactive species. It is also not known if ingestion of dietary antioxidants is directly related to development or prevention of degenerative diseases or how much is needed to afford protection (Food and Nutrition Board 2000).

8.5   Antioxidant Vitamins and Minerals

In addition to their conventional role in preventing deficiency diseases, a great number of nutrients are known to have antioxidant properties. Although scientists acknowledge the antioxidative content of thousands of foods, beverages, herbs, and spices (Carlsen et al. 2010), the US Food and Nutrition Board recognizes only three antioxidants, namely, vitamin C, vitamin E, and trace mineral selenium, though they also acknowledge that β-carotene influences biochemical reactions pertaining to the oxidative process (Food and Nutrition Board 2000). According to their definition, a dietary antioxidant is a substance found in human diets that significantly decreases the adverse effects of ROS and RNS on normal physiologic functions in humans. The criteria for this definition are as follows:

> The substance is found in human diets, the content of the substance has been measured in foods commonly consumed, and, the substance decreases the adverse effects of reactive species *in vivo* in humans (Food and Nutrition Board 2000, p. 17)

The following descriptions are taken from the Dietary Reference Intakes (DRIs) for antioxidants and related compounds (Food and Nutrition Board 2000) except, as otherwise, stated.

8.6   Vitamin E

There is a common perception that vitamin E is an age enhancer among the elderly. Vitamin E is a chain-breaking antioxidant in lipids and fat soluble, which prevents the propagation of free radical reactions which would otherwise cause lipid peroxidation of membranes and cellular lipoproteins. Specifically, it protects polyunsaturated fatty acids (PUFAs) from attack by peroxyl free radicals. This protection derives from the fact peroxyl radicals react a thousand times faster with vitamin E than with PUFAs.

Vitamin E deficiency is extremely rare in humans and is only associated with malabsorption of the vitamin (as in cystic fibrosis) or inborn errors in vitamin E metabolism. Experimental vitamin E deficiency is characterized by peripheral neuropathy and degeneration of axons of sensory neurons. Vitamin E supplements are sold as esters (to protect shelf life) of the natural form or as the synthetic mixture. When α-tocopherol is derived from vegetable oils, it is labeled as a natural source of vitamin E. Dietary sources for the vitamin are fats and oils, vegetables, meat, fish, poultry, and breakfast cereals. For adult men and women, the RDA for vitamin E is 12–15 mg/day. Upper limit is set at 1000 mg/day.

8.7 Vitamin C (Ascorbic Acid)

The ubiquitous vitamin C or ascorbic acid, for which late Linus Pauling won a second Nobel for chemistry, is the most talked about and discussed antioxidant in scientific literature, offering a wide spectrum of benefits to the human body, starting from protection against common cold right up to act as a check to the spread of the ubiquitous bacteria *E. coli*, which otherwise is beneficial but can turn virulent when the body's immunity falls, adding problems to a diabetic because the carbon (excess glucose in the blood) is a ready source of sustenance for the bacteria. It is a water-soluble, broad-based antioxidant which quenches a variety of ROS and RNS. Additionally, it can also regenerate or spare α-tocopherol. When this intercepts a free radical, α-tocopheroxyl radical is formed. This radical can be reduced by vitamin C (or other reducing agents), thereby oxidizing vitamin C and returning vitamin E to its reduced state. Thus, vitamin C has the capacity to recycle vitamin E.

Vitamin C is also a cofactor for enzymes involved in a variety of biological functions, including the synthesis of collagen and other connective tissues, the biosynthesis of carnitine for cellular energy production, and the interconversion of several major neurotransmitters, including dopamine and norepinephrine. Specifically, ascorbic acid is a cofactor for dopamine β-hydroxylase, which converts dopamine to norepinephrine, and is also involved in the hydroxylation of tryptophan in the brain to form serotonin (5-HT). Ascorbic acid has other functions in nervous tissues, as well. It modulates the activity of glutamatergic and dopaminergic neurons and is involved in the synthesis of glial cells and myelin. Vitamin C is highly concentrated in the CNS and local brain concentrations and changes rapidly with neuronal activity. Moreover, brain pools are relatively resistant to vitamin C depletion. Together, these observations suggest a major role for vitamin C in CNS functioning. The protective effects of vitamin C in the brain may arise from its free radical scavenging property or may be related to its actions on central vascular tissues. Which mechanism predominates, it is uncertain to say. In the periphery, vitamin C has vasodilatory and anticlotting effects and is thought to play a role in the reduction of cardiovascular disease by inhibiting plasma low-density lipoproteins (LDL) cholesterol oxidation. Oxidized LDL tends to aggregate on vascular cell walls resulting in the accumulation of plaques which narrow blood vessels. Since senile dementia and other neurodegenerative diseases may involve narrowing of cerebral blood vessels; vitamin C may serve similar functions in the brain.

Close to 90% of vitamin C comes from fruits and vegetables in the western diet. Major dietary contributors include citrus fruits and juices, green vegetables, tomatoes and tomato products, and potatoes. The RDA for vitamin C is 90 mg/day and 75 g/day for male and female adults, respectively. Food and supplement intake rarely exceeds 200 mg/day. Excess intake can lead to gastrointestinal disturbances, cramps, and diarrhea. The UL for vitamin C for adults is 2000 mg/day.

8.8 Selenium (Se)

It is of late that selenium (Se) has been recognized for its importance in the health for humans, animals, and plants. Selenium is a trace element which is required for human health, which functions as selenoproteins, of which two classes are known, the first one consisting of a family of selenium – dependent, glutathione peroxidase enzymes which serve as the body's primary defense mechanism against oxidative damages. Glutathione peroxidase is widely distributed in the body but is highly concentrated in the brain where it is localized in glial cells in central gray matter, hippocampus, and temporal cortex. In Alzheimer's and Parkinson's patients, a decreased activity of this enzyme has been noted which points to oxidative stress in them. The second class of selenium-dependent enzymes, the iodothyronine deiodinases, regulates the thyroid hormone metabolism. Given the importance of the thyroid hormones in CNS function, it is possible that even subclinical deficiencies of the element may play a role in psychological function (Benton 2002).

Selenium is ubiquitous in food products, such as meat, shellfish, dairy products, and plant foods. Selenium is found in soil, but its concentration can widely differ depending on the soil. Selenium content of the plant foods depends on its content in the soil, in which the plant is grown. Typical intakes in the USA and Canada from foods are well above the prescribed minimum of 55 microgram/day, and adverse effect in humans due to high intake is rare, and the UL for adults is 400 mg/day.

8.9 Vitamin D

Though vitamin D has not been technically recognized an antioxidant, during the last decade and a half, this vitamin has been shown to regulate adrenalin, noradrenaline, and dopamine, protecting against serotonin depletion and thereby having a possible biological relationship to depression. Epidemiological investigations verify that low levels of vitamin D are associated with depression (Hoang et al. 2011). However, recent investigations are skeptical about its effectiveness in relieving depression, and in one investigation it was shown to worsen depression (Spedding 2014).

8.10   Oxidative Damage and CNS Disorders

The brain's oxygen utilization rate is high *vis-à-vis* its high energy needs, which renders it to oxidative damages. The brain also has a high content of fatty acids incorporated into neural membranes, and much of these are PUFAs which are especially vulnerable to lipid peroxidation. Additionally, the brain has high levels of transition metals, such as copper and zinc, which readily form reactive hydroxyl species (Farris and Zhang 2003).

8.11   Role of Dietary Supplements in CNS-Related Cognitive Decline

8.11.1   Antioxidant Vitamins

Many human investigations have demonstrated the beneficial effects of vitamin E and vitamin C on neurodegenerative processes in older individuals. In patients with moderate symptoms of Alzheimer's disease, high doses of both vitamins slowed the progression to severe dementia and delayed the time to institutionalization of these patients by about 7 months (Sano et al. 1997). However, no differences were observed between treatment and controls in tests of cognitive functioning, possibly because of the advanced stage of the disease in these patients. Similarly, in a study of elderly individuals with mild cognitive impairment, subjects were randomly assigned to receive either 2000 IU of vitamin E daily or a placebo. After 3 years, no significant differences were shown between vitamin E and the placebo groups in their rate of progression to the disease (Petersen et al. 2005).

The relationship between antioxidants and neurodegenerative diseases has been examined through epidemiological studies, which is an exciting field of study. This research has shown that vitamin C and especially vitamin E may be associated with a lower risk of Alzheimer's disease. However controversy surrounds the question whether the benefits have been provided by the antioxidants present in the foods ingested per se or through the supplementary antioxidants. From a large study of population (5300 adults) above the age of 55 in Rotterdam, the Netherlands, and free of dementia at baseline, it was observed over a 6-year period that high dietary intake of vitamins C and E was strongly associated with lower risk of the disease, among smokers, but was less transparent among nonsmokers (Engelhart et al. 2002). One explanation for these findings could be that smoking is associated with lower plasma concentrations of the major antioxidants, and consequently, smokers may be less protected against the formation of free radicals and oxidative stress than nonsmokers. Surprisingly, intake of vitamin E from supplements (as opposed to diet) was not associated with risk reduction of the disease in the Rotterdam study cited above.

Another approach to assess the association between antioxidants and cognitive impairments is to measure serum concentrations of nutrients as markers of dietary behavior. Several studies have shown that serum concentrations of vitamins C, E, and A are lower in Alzheimer's patients than aged-matched controls without the disease (Grundman and Delaney 2002). Since serum investigations reflect dietary intake over time, these data imply that low habitual intake of these vitamins might have contributed to the development of the disease. A population-based study of elderly individuals living in Basel, Switzerland, on the relationship between antioxidant intake and memory performance over a 12-year period indicated that higher concentrations of plasma A and C, and not E, predicted better semantic memory in a vocabulary test but not other memory tasks such as free recall, recognition, and working memory.

Data on the effects of antioxidant vitamins on the progression of Parkinson's disease (PD) are also conflicting (Farris and Zhang 2003). For example, studies that assessed dietary intake of vitamin E suggest protective effect against PD (Etminan et al. 2005), with further supportive data from the investigation of Fahn (1991) which showed that high doses of vitamins C and E delayed the need for medication in patients with PD by 2–5 years, while another double-blind, placebo-controlled trial on vitamin E supplementation led to no benefits to the patients with PD. The study of Fahn (1991) was not blinded or controlled, and patients were permitted to take other medications concurrently with vitamin supplements.

8.11.2   Folate B_{12} and Homocysteine

There is the possibility of efficacy of vitamin B supplements in combating cognitive decline, as elevated homocysteine may be involved in the development of cognitive decline and dementia in the elderly. Folate and B_{12} are required for the conversion of homocysteine to the essential amino acid methionine. Lack of either vitamin reduces the activation of this pathway and leads to the accumulation of homocysteine in the blood. Homocysteine is thought to have direct toxic effects on CNS neurons by exacerbating oxidative stress or eliciting DNA damage. Elevated plasma homocysteine has also been implicated as a strong risk factor for the Alzheimer's disease (Le Boeuf 2003). High plasma homocysteine has also been observed in patients with PD (Kunhn et al. 1998). Elevated plasma homocysteine is typically associated with low and marginal intake of folate and vitamin B_{12}. The elderly are at greater risk of deficiencies of these nutrients due to reduced absorption of folate and vitamin B_{12} which naturally occurs with aging (Food and Nutrition Board 2000). Thus, it has been hypothesized that elderly individuals with lower concentrations of folate and B_{12} and higher concentration of homocysteine may have poorer cognitive functioning than elders with normal indices of these constituents.

Oral supplements of folate and B_{12} restore normal plasma indices of these nutrients and reduce homocysteine levels. Thus, one strategy would be to supplement the diets of older individuals with folate and B_{12} to correct the disturbance in homocysteine. There is no conclusive evidence on the positive effects of folate and B_{12} on correcting cognitive function, and a recent meta-analysis, encompassing 11 clinical trials and 22,000 participants, concluded that while B vitamins were successful in lowering levels of homocysteine, no significant effects were shown in maintaining cognitive function.

8.11.3 Herbs and Botanicals

For millennia, herbal medicine was the only source for human health, with written records testifying that their use dates back to at least 5000 years (Swerdlow 2000). In fact, it is estimated that as many as one-half of currently used drugs were originally derived from plants (Barrett et al. 1999). The best existing example is the ancient system of Indian medicine, known as *Ayurveda* (the word is derived from the ancient Indian language Sanskrit, which means the gospel of health, the last word in health). A herb can be any form of a plant, including its leaves, flowers, seeds, stems, or roots. A herbal supplement, in turn, may be the herb sold in its raw form or as an extract, where the plant is macerated with water or another solvent to extract some of its chemicals.

The following discussion will confine to the five most popular herbal supplements, though hundreds of them are available in the market. Following descriptions are borrowed from the reviews by Ernst (2002), except where specified. When the neuroprotective mechanism of an herb or botanical is known (or at least strongly suspected), it is described as well. The following is a description of some of the important herbs.

8.11.4 Ginkgo biloba and Dementia

Ginkgo biloba is a herb derived from the leaves and nuts of the gingko or popularly called "maiden hair" tree as shown in Fig. 8.1.

This herb has been used to treat asthma and chilblains (sores of the hands and feet from exposure to the cold) in Chinese medicine for millennia. Pharmacological studies suggest that this herb is anti-edemic, anti-hypoxic, antioxidant, anticoagulant, and a free radical scavenger (Ernst 2002). Put together, these actions could protect vascular tissues from ischemic damage (hypoxia). The herb has been experimentally used to protect against myocardial reperfusion injury, depression, brain trauma, memory impairment, dementia, and muscle pain. Extract of the herb contains the active ingredients flavonoid glycosides and terpene lactones (Ernst 2002).

Fig. 8.1 *Gingko biloba.*
(*Source:* Shutterstock)

The former may act as free radical scavengers, which could explain their action on brain trauma, cardiac injury, memory loss, and dementia. The latter may be responsible for antagonizing platelet clotting factors and increasing blood fluidity. Standardized extracts of the herb (designated as EGb 761) contain 22–27% flavonoid glycosides and 5–7% terpene lactones (consisting of ginkgolides A, B, and C and bilobalide).

The herb may provide neuroprotection by several mechanisms of action. EGb 761 has positive effects on cholinergic, dopaminergic, and serotonergic systems. For example, in one investigation, EGb 761 was shown to block the age-related decline in the density of acetylcholine and 5-HT 1A receptors in the brains of aged rats. In a variety of investigations, EGb 761 was shown to reduce the production of hydroxyl and peroxyl radicals, increased the activity of antioxidant enzyme systems, and inhibited oxidative damage to mitochondria. The latter activity is of major importance in aging since oxidative damage to mitochondrial DNA is a major target of free radical attack.

A large-scale randomized controlled trial, called the Ginkgo Evaluation of Memory (GEM) investigation (Snitz et al. 2009), was conducted to understand whether ginkgo could prevent cognitive decline in healthy, elderly individuals. However, no differences were observed between treatment and placebo on a battery of neuropsychological tests including memory, attention, visuospatial abilities, language, or executive function. On the basis of these results and others, it can be concluded that ginkgo extract provides positive therapeutic benefits to patients with cerebral disorders but little or no improvement in mental functioning for healthy individuals. Some caution is in order, however, as gingko use has been associated with spontaneous bleeding (Bent et al. 2005).

8.11.5   *Ginseng and Energy*

Roots of Asian ginseng (*Panax ginseng*) are believed to have sedative, hypnotic, and enlightening properties. The plant's (Fig. 8.2) extract also acts as a CNS stimulant and potentiates the stimulatory effects of caffeine from coffee, tea, and cola. The herb is used in traditional Chinese medicine to invigorate the body, restore sexual virility, and cure nervous disorders, among therapeutic benefits (Lieberman 2001).

The plant extract has been tested as a therapeutic agent for improving memory and mood. *Panax ginseng* has often been confused with Russian, American, and Japanese ginseng. The discussion here refers to *Panax ginseng*, unless otherwise noted.

The plant is likely to produce a number of side effects, such as insomnia, nausea, diarrhea, and headache (Ernst 2002). It also lowers blood glucose. Thus, the use of this herb might be counter-indicated in individuals taking antidiabetic drugs. It boosts energy levels in the body and relieves stress and fatigue. In early studies, the use of ginseng seemed to produce a number of beneficial effects, including aerobic capacity and blood lactate (Forgo and Schimert 1985). However, more recent studies using double-blind procedures have failed to find effects on physical performance, ratings of perceived exertion, or physiological parameters such as heart rate. A systematic review of 16 randomized, placebo-controlled trials found no compelling evidence that ginseng had any positive effects on physical or cognitive performance, let alone any specific disease state (Vogler et al. 1999).

Fig. 8.2 Asian ginseng
(*Panax Ginseng*). (*Source:*
Shutterstock)

8.11.6   St. John's Wort and Depression

St. John's wort (SJW, *Hypericum perforatum*) is a wild-growing herb with yellow flowers (Fig. 8.3, below), which has been used since ancient times to treat mental disorders and nerve pain.

When applied topically, as a balm, it was used to treat insect bites, wounds, and burns. To date St. John's wort (SJW) is used primarily to treat mild to moderate depression. However, it is not effective in treating major depression. The main active constituents of SJW are hypericin and hyperforin, although other components may be active, as well. The mechanism of action against depression is uncertain, but at least two pathways have been suggested. First, brain serotonin levels are low in depression. The herb may act by selectively inhibiting reuptake of serotonin (and perhaps other monoamines, Ernst 2002). Additional evidence is that SJW enhances monoamine metabolism which comes from studies showing that SJW administered in the prefrontal cortex of awake rats increased 5-HT and especially dopamine turnover in these animals. SJW has fewer side effects than conventional antidepressants, which makes it an attractive treatment alternative. Although the use of SJW has been associated with dry mouth, dizziness, gastrointestinal effects, increased sensitivity to light, and fatigue, the frequency of such complaints is quite low (Ernst 2002). However, other more serious side effects have been observed with the use of SJW, such as rapid deactivation of several classes of drugs, by inducing liver-detoxifying enzymes. Serious interactions are known to occur with protease inhibitors used to treat HIV infection, immunosuppressant drugs, birth control pills, cholesterol-lowering drugs, cancer, antiseizure medications, and blood anticoagulants (Izzo 2005). Thus, SJW is considered safe when taken alone but can lead to significant health problems when combined with other medications.

Fig. 8.3 St. John's wort
(*Hypericum perforatum*).
(*Source:* Shutterstock)

It should be noted that one of the more successful herbal therapies is the use of SJW in the treatment of mild to moderate depression. The general conclusion from more than 40 randomized clinical trials on the efficacy of SJW on depression is that it was more effective than placebo and comparable to tricyclic antidepressants to improve depressive symptoms. However, it is notable that two large, multicenter trials performed in the USA on SJW failed to show a therapeutic benefit on major depressive symptoms (Hypericum depression Trial Study Group 2002).

8.11.7 Kava: The Antidote for Anxiety

Depression and anxiety are two of the most serious mental disorders of modern times. There are many herbals used in Ayurveda to counteract these disorders in human treatment. Kava is one such. It is made from the dried rhizome (underground stem) of the plant *Piper methysticum*, which was traditionally used as a recreational drink in the South Pacific. A diagrammatic representation is shown in Fig. 8.4.

Kava has anxiolytic properties and also acts as a muscle relaxant, mood enhancer, analgesic, and sedative. To date it is generally used to treat seizures and psychotic illnesses (Ernst 2002). The active compounds are a family of kavapyrones, the anxiolytic actions, which are complex. The kavapyrones enhance gamma-aminobutyric acid (GABA) binding in the amygdale and inhibit norepinephrine uptake. Their muscle-relaxing effects are due to inhibition of Na and Ca channels as well as effects on the glutamate system. Kava potentiates the effects of other anxiolytics and alcohol and so should be avoided by individuals taking psychotropic medications. Long-term use has been associated with yellow discoloration of the skin, hair, and nails, visual disturbances, dizziness, ataxia, and hair, hearing, weight, and

Fig. 8.4 Kava *Piper methysticum. (Source:* Shutterstock)

appetite loss. Most disturbing are reports that Kava may induce toxic liver damage. The reports led the FDA to issue a consumer warning on Kava regarding this risk (US Food and Drug Administration 2002).

Kava has been shown to be beneficial in the treatment of anxiety disorders. In several double-blind, placebo-controlled trails, Kava was more effective in reducing anxiety than placebo and comparable to standard anxiolytic drugs (Fugh-Berman and Cott 1999). A meta-analysis, which used total score on the Hamilton Anxiety Scale as the common outcome measure, showed a 5-point reduction in the anxiety score with Kava, as compared to placebo (Pittler and Ernst 2003). Despite the beneficial effects of Kava, questions regarding its safety profile still remain. However, the sum of trials to date suggests that dosages less than 400 mg/day do not cause serious side effects and that liver toxicity is an extremely rare occurrence.

8.11.8   Valerian and Insomnia

Valerian (*Valeriana officinalis*) is a flowering plant which throws up sweet-scented flowers which bloom in the summer months. The usable product is prepared from roots and underground stems of the plant. It was Hippocrates who first described its properties as a medicinal plant, and, in the second century, the physician Galen prescribed it as a remedy for insomnia. In medieval Sweden, it was sometimes placed in the wedding clothes of the groom to ward of the "envy" of the elves (Thorpe 2013). During the sixteenth century, valerian flower extracts were used as a perfume, but its sedative effects were recognized by Europeans as early as the eighteenth century. The plant is pictorially shown in Fig. 8.5.

The plant extract acts as an inhibitor of sympathetic nervous system neurons by modifying the transport and liberation of the GABA neurotransmitter. Although the mechanism of action of valerian as a mild sedative is not fully understood, it is generally believed that some of the GABA analogs, in particular, valerenic acids, appear to have some affinity for the GABA receptor, a class of receptors on which benzodiazepines are known to act (Hozl and Godau 1989). Although benzodiazepines are the drugs deemed most effective in combating insomnia, their prolonged use can produce a number of adverse side effects. Although valerian may also cause mild side effects, such as prolonged tiredness, dizziness, headaches, or upset stomach, it is believed to be generally safe for use over short periods of time (Taibi et al. 2007).

The bulk of evidence on valerian as a sleep aid is controversial, in spite of the dozens of investigations on the product. Though a meta-analysis of 16 randomized, placebo-controlled trails concluded that valerian might improve sleep, some evidence of publication bias was noted in the measurement of sleep quality (Bent et al. 2006). A concomitant review of 29 controlled trials concluded that most studies found no significant differences between valerian and placebo, with all of the more recent and methodologically rigorous studies failing to do so (Taibi et al. 2007).

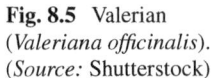

Fig. 8.5 Valerian
(*Valeriana officinalis*).
(*Source:* Shutterstock)

8.11.9   Fish Oil Supplements

The very high scale of interest in fish oils, as health supplements, is a, rather, recent phenomenon. Even those who are vegetarians are veering toward them. In the USA, there has been a veritable explosion in the sale of fish oil and omega-3 supplements, which saw an increase of 145% between 2007 and 2012, aggregating to a sale of over US $ 1 billion (Grey and Bolland 2014), which was purchased by 67% of the population which consumed dietary supplements (Consumer Lab.com 2010). *In* the developing world, the trend is still to catch up on a scale which is comparable to that seen in the USA, but great interest is unfolding. In the USA, the reason for this is because the American Heart Association endorsed their use in the secondary prevention of heart disease. Given their importance in brain development and functioning (Wainwright 2002), fish oil supplements, namely, eicosapentaenoic acid (EPA) and docosahexaenoic acid (DHA), have been thought to play a crucial role in proper brain function.

8.12   The Interrelationship Between Fish Oil Supplements and Cognitive Decline

There is largely supportive evidence from epidemiological studies that show fish intake to be positively linked to enhanced cognitive capacity. The late Sir Arthur Conan Doyle creator of the famed Sherlock Holmes-Dr Watson duo, who solved several murder mysteries, has credited the brainy Sherlock Holmes, as a voracious fish eater. Cunnane et al. (2009) reviewed observational studies that related fish or dietary DHA/EPA consumption to the risk of cognitive decline or dementia and indicated that 9 out of 12 studies showed at least some protective effects on Alzheimer's disease, cognitive decline, or all-cause dementia. Whether DHA/EPA supplements will also exert a protective effect on cognition performance has been examined in at least half a dozen investigations since 2000. Providing DHA + EPA supplements for 6 months (Suzuki et al. 2001) and using DHA (+arachidonic acid in unspecified proportions), Kotani et al. (2006) showed better performance on tests of cognitive abilities in subjects who had exhibited mild cognitive impairment. In light of the vast numbers of health conscious individuals, who currently take fish oil supplements, in many instances to ward off future cognitive problems, including dementia and/or Alzheimer's disease, it is worth asking if Omega-3 fatty acids are effective or not in preventing these disorders in later life, who are currently healthy without any cognitive disorders. A recent review identified three high-quality randomized control trials that used DHA + EPA supplements anywhere from 6 to 40 months (Sydenham et al. 2012), with over 3500 individuals comprising the total subject pool across the investigations. The reviewers concluded that there was no benefit to cognitive function from the omega-3 fatty acid supplements in cognitively healthy individuals over the age of 60; instead, weekly consumption of oily fish was recommended by the review's authors.

8.13   The Interrelationship Between Fish Oil and Psychiatric Disorders

Additional to what has been discussed above, researchers have also investigated whether omega-3 fatty acid can combat mental health problems. A few studies with depressed patients deemed as treatment-nonresponsive have found EPA supplements to be more effective than placebos when added to their existing antidepressant treatment (Su et al. 2003). Work by Peet and Horrobin (2002) is instructive, as their dose-response design showed a 1 g/day dose to be effective, in contrast to 2 or 4 g/day dosages (2 g/day being the norm). Schizophrenia, a severe mental disorder characterized by a breakdown in thinking and accompanied by poor emotional responses, anxiety, and depression, has also been the subject of investigations using omega-3 fatty acids. Although Fenten et al. (2001) found no therapeutic benefit from EPA treatment in a placebo-controlled trial, the results from other investigations are promising. For example, Peet et al. (2001) treated schizophrenic patients with either placebo or EPA as a monotherapy for 3 months, and at the end of which, all the 12 patients in the placebo group required treatment with antipsychotic drugs, in contrast to only 6 of the 14 patients receiving EPA. In addition, symptom ratings were significantly lower for the EPA group than the placebo group.

References

Barrett, B., Kiefer, D., & Rabago, D. (1999). Assessing the risks and benefits of herbal medicine: An overview of scientific evidence. *Alternative Therapies in Health and Medicine, 5,* 40–49.

Bent, S., Goldberg, H., Padula, A., & Avins, A. L. (2005). Spontaneous bleeding associated with Ginkgo biloba: A case report and systematic review of the literature. *Journal of General Internal Medicine, 20,* 657–661.

Bent, S., Padula, A., Moore, D., Patterson, M., & Mehling, W. (2006). Valerian for sleep: A systematic review and meta-analysis. *American Journal of Medicine, 119,* 1005–1012.

Benton, D. (2002). Selenium intake, mood and other aspects of psychological functioning. *Nutritional Neuroscience, 5,* 363–374.

Carlsen, M. H., Halvorsen, B. L., Holte, K., Bohn, S. K., & Draglund, S. (2010). The total antioxidant content of more than 3100 foods, beverages, spices, herbs and supplements used worldwide. *Nutrition Journal, 9,* 3.

Cohen, P. A. (2012). DMAA as a dietary supplement ingredient. *Archives of Internal Medicine, 172,* 1038–1039.

ConsumerLab. com. (2010). *Survey of vitamin and supplement users.* Available at: http://www.consumerlab.com/news/Supplement_Survey_Report/1_31_2010. Accessed 25 Mar 2015

Cunnane, S. C., Plourde, M., Pifferi, F., Begin, M., Feart, C., & Barberger-Gateau, P. (2009). Fish, docosahexaenoic acid and Alzheimer's disease. *Progress in Lipid Research, 48,* 239–256.

Engelhart, M. J., Geelings, M. I., Ruitenberg, A., van Swieten, J. C., Hofman, A., Witteman, J. C., & Breteler, M. M. (2002). Dietary intake of antioxidants and risk of Alzheimer disease. *Journal of the American Medical Association, 287,* 3223–3229.

Ernst, E. (2002). The risk-benefit profile of commonly used herbal therapies: Ginkgo, St John's Wort, Ginseng, Echinacea, Saw Palmetto, and Kava. *Annals of Internal Medicine, 136,* 42–53.

Etminan, M., Gill, S. S., & Samii, A. (2005). Intake of vitamin E, vitamin C, and carotenoids and the risk of Parkinson's disease: A meta-analysis. *The Lancet Neurology, 4,* 362–365.

Fahn, S. (1991). An open trial of high-dosage antioxidants in early Parkinson's disease. *American Journal of Clinical Nutrition, 53,* 380S–382S.

Farris, M. W., & Zhang, J. G. (2003). Vitamin E therapy in Parkinson's disease. *Toxicology, 189,* 129–146.

Fenten, W. S., Dickerson, F., Boronow, J., Hibbeln, J. R., & Knable, M. (2001). A placebo-controlled trial of omega 3-fatty acid (ethyl eicosapentaenoic acid) supplementation for residual symptoms and cognitive impairment in schizophrenia. *American Journal of Psychiatry, 158,* 2071–2074.

Food and Nutrition Board, Institute of Medicine and National Academy of Sciences. (2000). *Dietary reference intakes for vitamin C, vitamin E, selenium and carotinoids.* Washington, DC: National Academy Press.

Forgo, I., & Schimert, G. (1985). The duration of effect of the standardized ginseng extract G 115 in healthy competitive athletes. *Notabene Medici, 15,* 636–640.

Fugh-Berman, A., & Cott, J. M. (1999). Dietary supplements and natural products as psychotherapeutic agents. *Psychosomatic Medicine, 61*, 712–728.

Grey, A., & Bolland, M. (2014). Research letter: Clinical trial evidence and use of fish oil supplements. *JAMA Internal Medicine, 174*, 460–462.

Grundman, M., & Delaney, P. (2002). Antioxidant strategies for Alzheimer's disease. *Proceedings of the Nutrition Society, 61*, 191–202.

Hoang, M. T., DeFina, L. F., & Willis, B. L. (2011). Association between low serum 25-hydroxyvitamin D and depression in a large sample of healthy adults: The cooper center longitudinal study. *Mayo Clinic Proceedings, 86*, 1050–1055.

Hozl, J., & Godau, P. (1989). Receptor binding studies and *Valeriana officinalis* on the benzodiazepine receptor. *Planta Medica, 55*, 642.

Hypericum Depression Trial Study Group. (2002). Effect of *Hypericum perforatum* (St John's wort) in major depressive disorder: A randomized controlled trial. *Journal of the American Medical Association, 287*, 1807–1814.

Izzo, A. (2005). Herb-drug interactions: An overview of the critical evidence. *Fundamentals of Clinical Pharmacology, 19*, 1–16.

Kotani, S., Sakaguchi, E., Warashina, S., Matsukawa, N., & Ishikura, Y. (2006). Dietary supplementation of arachidonic and docosahexaenoic acids improves cognitive dysfunction. *Neuroscience research, 56*, 159–164.

Kunhn, W., Roebroek, R., Blom, H., van Oppenraaij, D., Przuntek, H., Kretschmer, A., Buttner, T., Woitalla, D., & Muller, T. (1998). Elevated plasma levels of homocysteine in Parkinson's didease. *European Neurology, 40*, 225–227.

Le Boeuf, R. (2003). Homocysteine and Alzheimer's disease. *Journal of the American Dietetic Association, 103*, 304–307.

Lieberman, H. R. (2001). The effects of ginseng, ephedrine, and caffeine on cognitive performance, mood and energy. *Nutrition Reviews, 59*, 91–102.

Peet, M., & Horrobin, D. F. (2002). A close-ranging study of the effects of ethyl-eicosapentaenoate in patients with ongoing depression despite apparently adequate treatment with standard drugs. *Archives of General Psychiatry, 59*, 913–919.

Peet, M., Brind, J., Ramchand, C. N., Shash, S., & Vankar, G. K. (2001). Two double-blind placebo-controlled pilot studies of eicosapentaenoic acid in the treatment of schizophrenia. *Schizophrenia Research, 49*, 243–251.

Petersen, R. C., Thomas, R. G., Grundman, M., Bennett, D., & Doody, R. (2005). Vitamin E and donepezil for the treatment of mild cognitive impairment. *New England Journal of Medicine, 352*, 2379–2388.

Pittler, M. H., & Ernst, E. (2003). Kava extract for treating anxiety. *The Cochrane Database of Systematic Reviews, 1*, CD003383. https://doi.org/10.1002/14651858.CD003383.

Sano, M., Ernesto, C., Thomas, R. G., Klauber, M. R., & Schafer, K. (1997). A controlled trial of selegiline, alpha-tocopherol, or both as treatment for Alzheimer's disease. The Alzheimer's disease cooperative study. *New England Journal of Medicine, 336*, 1216–1222.

Snitz, B. E., O'meara, E. S., Carlson, M. C., Arnold, A. M., & Ives, D. G. (2009). *Ginkgo biloba* for preventing cognitive decline in older adults: A randomized trial. *Journal of the American Medical Association, 302*, 2663–2670.

Spedding, S. (2014). Vitamin D and depression: A systematic review and meta-analysis comparing studies with and without biological flaws. *Nutrients, 6*, 1501–1518.

Su, K.-P., Huang, S.-H., Chiu, C.-C., & Shen, W. W. (2003). Omega-3 fatty acids in major depressive disorder: A preliminary double-blind, placebo-controlled trial. *European Neuropsychopharmacology, 13*, 267–271.

Suzuki, H., Morikawa, Y., & Takahashi, H. (2001). Effect of DHA oil supplementation on intelligence and visual acuity in the elderly. *World Review of Nutrition and Dietetics, 88*, 68–71.

Swerdlow, J. L. (2000). Modern science embraces medicinal plants. In J. L. Swerdlow (Ed.), *Nature's medicine: A chronicle of mankind's search for healing plants through the ages* (pp. 110–157). Washington, DC: National Geographic Society.

Sydenham, E., Dangour, A. D., & Lim, W. S. (2012). Omega 3 fatty acid for the prevention of cognitive decline and dementia. *Cochrane Database of Systematic Reviews, 6*, CD005379. https://doi.org/10.1002/14651858.CD005379.pub3.

Taibi, D. M., Landis, C. A., Petry, H., & Vitiello, M. V. (2007). A systematic review of valerian as a sleep aid: Safe but not effective. *Sleep Medicine Reviews, 11*, 209–230.

Thorpe, B. (2013). *Northern mythology, comprising the principal popular traditions and superstitions of Scandinavia, northern Germany and the Netherlands* (Vol. 2). London: Forgotten Books. (Original work published 1851).

US Food and Drug Administration. (1994). *Dietary supplement and health education act of 1994.* Available at: http://www.fda.gov/Regulatory Information/Legislation/FederalFoodDrugand CosmeticActFDCAct/Significant Amendments to the FDCA Act/ucm 1 480003.htm. Accessed 30 June 2015

US Food and Drug Administration, Center for Food Safety and Applied Nutrition. (2002). *Consumer advisory: Kava-containing dietary supplements may be associated with severe liver damage.* Available at: http://www.fda.Gov/foodresourcesforyou/consumers/ucm085482.htm. Accessed 25 Mar 2015

Vogler, B. K., Pittler, M. H., & Ernst, E. (1999). The efficacy of ginseng. A systematic review of randomized clinical trials. *European Journal of Clinical Pharmacology, 55*, 567–575.

Wainwright, P. E. (2002). Dietary essential fatty acids and brain function: A developmental perspective on mechanisms. *Proceedings of the Nutrition Society, 61*, 61–69.

Chapter 9
The Sugar: Behavior Link

Abstract The chapter would principally discuss the "meaning" of sugar, its biochemistry and implications in human diet, the sugar metabolism, reasons why we veer toward the sweet taste, the interrelationship between sugar intake and cognitive behavior, sugar-affected mood fluctuations, sugar and hyperactivity, interactions between sugar, drug, and alcohol. Additionally, the chapter would also discuss the role of sugar in controlling pain.

Keywords Sugar · Behavior · Sugar metabolism · Sweet taste · Hyperactivity · Pain

Sugar is an enigma of human life. The "love-hate" relationship with sugar and humans is well-known, the world over. That is what makes its study in relation to human behavior worth the while. When we were children, we were told "a spoonful of sugar helps the medicines go down," and that "little girls are made of sugar and spice, and, everything is nice." As adults, we may call all those, whom we deeply care for, as "sweetheart," "honey," or "sugar pie" if they are females. In Indian languages there are no equivalents, especially where one's lady love is concerned, though in the widely spoken Indian language Hindi, the reference to one's lady love is *Mere Jaan*, meaning "my life." The equivalent translations in the regional languages are current, for example, *ente jeevan* (my life) in Malayalam, down south in Kerala, and *ennudai jeevan* in Tamil, in Tamil Nadu. On holidays, such as Valentine Day and Halloween, or even birthdays and anniversaries, we celebrate the event with gifts of candy, cake, and other sweets, where the "birthday cake" is the attraction of the day. In the discussion that follows, the meaning of the word "sugar," when used by nutritionists, forms the core.

9.1   The Meaning of "Sugar": Its Biochemistry and Implications in Human Diet

Sugars belong to the group of foods, known as carbohydrates, which are composed of the elements carbon, hydrogen, and oxygen. There are hundreds of different carbohydrates; however the discussion confines to only those which have a relevance to nutrition. These carbohydrates are classified as sugars, starches, and fibers. When one refers to the sugars in the diet, there is a reference to more than the granulated white powder in bowls on kitchen counters. Dietary sugars can be divided into the following three groups:

The simple sugars or monosaccharides which are the building blocks for all carbohydrates. The general formula for monosaccharides is $(CH_2O)n$, where n is three or greater. Monosaccharides can have as few as three or as many as nine carbon atoms. However, the one most commonly encountered in human diet has six carbon atoms, which is the glucose ($C_6H_{12}O_6$). Although glucose is not a major component of foods, it is the primary breakdown product of more complex dietary carbohydrates. Fructose, the sweetest of the monosaccharides, is found in fruits, in honey, and now in the form of high fructose corn syrup which is commonly used as a sweetener in soft drinks and other foods. Galactose, the third of the nutritionally relevant monosaccharides, is seldom found free in foods but is a component of the disaccharide lactose.

Disaccharides, which are made of two monosaccharides, is the main form of sugar of most of our sugar intake. Americans consume most of their sugar in the form of sucrose, a disaccharide comprised of one molecule of glucose and another of fructose. Sucrose, a very sweet carbohydrate, is what we find on our tables, and what most people mean when they say sugar. Sucrose is produced commercially from sugar beets and also sugarcane, which is also found in molasses, maple syrup, and small quantities in fruits and vegetables. Another important disaccharide is lactose, which contains one molecule of glucose and one of galactose, which is found in milk products. Finally, maltose is a disaccharide containing two molecules of glucose. Maltose does not occur naturally in food to any great extent.

The less familiar oligosaccharides are made up of three or more monosaccharide units, such as raffinose, a trisaccharide found in sugar beets and cabbage, or maltodextrin, a polysacchartide produced from starch. These are not the common compounds of food in human diet.

9.2   The Sugar Metabolism

The preliminary step in sugar metabolism is the breakdown of sucrose in the small intestine into its monosaccharide components, fructose and glucose, by an enzyme called sucrase. After absorption from the gastrointestinal tract, glucose is carried in the blood stream to the liver, brain, and other tissues. In the liver, glucose is removed from the blood and is stored in the form of glycogen. The liver can store enough glycogen to see us through a 10-h fast. Any excess that the liver cannot store is converted into fat and stockpiled in fat cells. Other enzymatic processes in the liver convert fats and amino acids into glucose, as well. Thus we can maintain blood glucose when eating a variety of macronutrients. When needed, glycogen is removed from the liver and broken to glucose by glucagon. Glucose is the primary fuel for the brain but is not stored in the brain (Saril 2001). Further, the brain lacks the enzymes which are present in the liver for converting amino acids and fats into glucose. Thus, the brain is dependent on the circulating blood for glucose, as its fuel, and experiences consequences related to fluctuations in glucose levels.

The metabolic pathway for fructose is somewhat different compared to that of glucose. In particular, fructose is absorbed from the gastrointestinal tract by a different mechanism than that for glucose. Following absorption from small intestine, it is transported to the liver where it is metabolized. Much of the fructose in the liver is converted to glucose. However, experimental data suggest that in addition to being converted to glucose, fructose can stimulate the production of fat in the body. This feature of fructose metabolism coupled with the increasing use of high-fructose corn syrup in sweetened beverages has led to a concern about the role of fructose in the currently encountered obesity epidemic in the USA.

9.3   Why Do We Veer Toward the Sweet Taste?

Most animals, including humans, have a strong preference for sweet-tasting substances. Ancient cave paintings by prehistoric artists provide evidence of the high esteem in which sweet-tasting foods, such as honey, figs, and dates, were held. Paintings from tombs, dating as early as 2500 BCE, depict the beekeeping practices of ancient Egyptians (Darby et al. 1977). There are chemeo receptors for four basic flavors, namely, salt, sour, bitter, and sweet, as well as a "savory" taste known as *umami*. When one eats something sweet, taste receptor cells on the tongue are activated and communicate via neurotransmission to the brain. The complete cascade of events which occurs between initial stimulation of one's taste buds and the final perception of sweet taste has yet to be completely understood. It is thought, however, that sweet taste is communicated to the brain via at least two means of transmission: via a G-protein second messenger system and via Ca- and K-gated channels (Lindemann 2001). Recent research investigating transduction of sweet taste uncovered the family of G-protein-couple sweet receptors, T1R1, T1R2, and T1R3. The T1Rs, when expressed, stimulate the generation of the receptor which binds sugars and sweet-tasting molecules (Lindemann 2001).

Although the nature of the hedonistic response to the sweet taste is not completely understood, there is considerable evidence supporting the actions of certain neurotransmitters in this response. The overwhelming evidence suggests that the hedonistic response is triggered in the brain, not in the periphery. Both the endogenous opioid system and the dopamine system may be involved. The former is similar to the response in which drugs like morphine and heroin act to alleviate pain. Merely eating or anticipation to eat sweet foods can cause the release of beta-endorphin in the brain, producing a rewarding feeling resulting from direct actions at endogenous opioid receptors via indirect or direct actions in the dopamine system. Finally, the two neurotransmitter systems can be separated into distinct categories such that dopamine seems to be involved in "wanting" palatable foods (such as sweet ones) and the endogenous opioid system is more involved in "liking" the food.

9.4 The Interrelationship Between Sugar Intake and Cognitive Behavior

The positive effects of acute sugar consumption on cognition have been demonstrated in all age groups, from infant to the very elderly, as well as the Down syndrome patients and those with Alzheimer's disease. As with infants, sugar intake can have positive effects on cognitive behavior in young children. For example, 9–12-year-old boys performed significantly better on a vigilance task shortly after they had consumed sugar-containing confectionery product, than when given a sweet tasting, but non-nutritive, placebo. While sugar intake did improve performance on the vigilance task, intake did not alter the children's ability to complete tasks measuring memory or visual perception. Similar positive effects on cognitive behavior have been observed after glucose ingestion in adult population.

Given what we know about the relationship between fasting and blood glucose (namely, if you go for a long-time fasting, your blood glucose is used by your body and thus there is less in your blood), fast duration is an important experimental consideration when evaluating the effects of glucose on cognitive tasks. Sunram-Lea et al. examined the difference between a short (2 h) fast and a long (9–12 h) fast on a variety of cognitive tests. Blood glucose for individuals fasting overnight was lower than for those who had fasted for only 2 h. Blood glucose was elevated following glucose ingestion but was still lower in those who fasted overnight relative to those who fasted only for 2 h. However, although individuals had different levels of blood glucose, all of them performed better on the battery of cognitive tests when they had consumed glucose than when they were given a placebo containing the non-nutritive sweetener aspartame. The findings in this experiment are important in that they studied the effects of glucose in a more naturalistic manner. People typically eat every few hours that they are awake, and positive effects of sugars on cognition can be produced under these conditions. In conclusion, intake of glucose and other sugars shortly preceding a mental task has been related to improvements in cognitive behavior in children, young adults, and the elderly. However, there is much we need to know about the effects of acute glucose intake on cognitive behavior. For instance, additional information is needed about the quantity of the sugar that should be consumed to enhance cognitive performance. While intake of a moderate amount of glucose may improve an individual's ability to attend a mental task, intake of a larger amount may have less beneficial effects.

9.5 Sugar Affected Mood Fluctuations

Indeed, our cultural attitude points to the belief that sugar intake, in any form, elevates human moods. People with depression and anxiety often report "self-medication" with palatable high-sugar/high-fat foods. People who reported greater levels of anxiety and depression and fatigue also reported greater cravings for high-carbohydrate/high-fat foods than people who craved protein-rich foods. Participants in this study, who classified themselves as "carbohydrate cravers," were given questionnaires asking what type of foods they typically craved and scales assessing mood prior to and following eating the foods, they reported craving the most (Christensen and Pettijohn 2001). The majority of the participants reported that eating the foods which they craved alleviated these negative moods. A corollary argument that links carbohydrate to mood fluctuation is that carbohydrate intake kicks up blood glucose, which in turn elevates mood. To some extent, research supports this argument (Benton 2003). Alternatively, some have noted that sugar consumption correlates with depression (Westover and Marangell 2002), but other research has not supported an effect of sugar intake to mood enhancement (Guo et al. 2014).

In research which examined sweet rewards, Willner et al. (1998) showed that humans in an experimental model of depression increased their craving for a sweet, namely, chocolate, versus less palatable carbohydrates. Turning more directly to sugar, Reid and Hammersley (1995) investigated the ability of sucrose to alter mood. In their first study, after a 9-h fast, participants were asked to complete the Profile of Mood States (POMS) at the start of the study and at subsequent time points. To avoid any influence of taste on mood, all subjects were given a benzocaine anesthetic lozenge prior to drinking one of the beverages: a sugar-sweetened drink, a saccharin-sweetened drink, or plain mineral water. There were no differences as a function of drink condition on any mood measures with the exception of elevation of energy in the sucrose condition in women who ordinarily ate breakfast. There were no differences for other women or for any of the men. It seems likely that in this experiment, the effects of mood related to sugar were attributable to the consumption of calories in people experiencing an abnormal caloric deprivation. Additionally, it is possible that the benzocaine lozenge itself contributed to the results of this study, since the lozenge is unpalatable and produces numbness of the mouth, tongue, and upper throat. In a subsequent study, the same researchers examined the effects of diet on mood in non-fasted participants. In this study, participants were asked to eat yogurt containing added corn oil, sugar, both corn oil and sugar, or saccharin and then completed a variety of scales which determine mood, hedonic response, fullness, and thirst (Reid and Hammersley 1999). The results from this study differ from those of the first reported earlier, above, in two ways. First, intake of sucrose did not increase subjective feelings of energy. This may be due, in part, to the fact that participants were not food deprived in this study. Second, intake of sugar and corn oil or sugar alone significantly increased self-ratings of calmness on a scale ranging from "calmer" to "more angry."

It seems that sugar has a subtle, yet demonstrable, post-ingestive effect on mood. However, these effects can be profoundly different depending upon the fasting state of the individual and the nature of the psychological test.

9.6 Sugar and Hyperactivity

A commonly held myth surrounding sucrose is that of its role in hyperactivity. Many parents and teachers believe that intake of sugary foods leads to an increase in activity in children and, specifically, an aggravation of attention deficit hyperactivity (ADHD). Over the years, hyperactivity and ADHD in children have drawn considerable attention from the medical and lay communities. In the USA, rates of ADHD as a diagnosis increased significantly over the past three decades (Centres For Disease Control and Prevention 2013). ADHD presents itself with a constellation of symptoms ranging from inattention to the stereotypical restlessness.

In 1994 a comprehensive investigation was conducted with a double-blind control trial to examine the possible negative effects of sucrose as well as aspartame on children's behavior and cognitive performance (Wolraich et al. 1994). The investigators supplied every member of the participating families with all of their foods and beverages for 3 weeks, with the weekly diets either high in sucrose without artificial sweeteners, low in sucrose but sweetened with aspartame, or low in sucrose but sweetened with saccharin (Wolraich et al. 1994). A large number of cognitive and behavioral measures were administered each week to both preschool age children and school-age children whose parents were described as being sensitive to sugar. For the preschool, of the 39 measures, only 2 significant differences arose across dietary treatments: children's performance on a pegboard task was slower during the sucrose condition and parental ratings of the children's cognition were better during the sucrose diet. For the school-age children, not 1 of 39 measures showed a difference over the 3 dietary periods. Relative to ADHD symptoms, no differences in conduct, attention deficit, hyperactivity, or oppositional behavior were found between the sucrose and non-sucrose diets (Wolraich et al. 1994). A meta-analysis subsequently executed by Wolraich et al. examined the findings of 23 studies on sugar intake and behavior in children. They only included experiments in which sugar compared to a placebo condition, where subjects were given aspartame or saccharin, and the experimenters were blind to the diet manipulation. Subjects were children with or without ADHD, and the experiments analyzed the effects of sugar intake on motor skills, like aggression, and mood. The ultimate conclusion of the meta-analysis was that sugar consumption had little or no effects on behavior, a conclusion reiterated by more recent reviews (Bellisle 2004).

9.7   Interactions Between Sugar, Drug, and Alcohol

Kanarek et al. (2005) examined the interactions between sucrose and the abuse of drugs, in both humans and animals. In experimental animals, intake of sugar-containing foods can predict subsequent drug intake, alter the rewarding properties (Vitale et al. 2003), and increase the pain-relieving properties of drugs, such as morphine (Kanarek et al. 2001). Additionally, preliminary evidence that sugar may be addictive comes from rodent models in which access to food cycles between 12 h of no food and 12 h of a sugar solution with a laboratory chow was allowed. After a month of this feeding schedule, rats displayed a number of behaviors similar to those observed in drug-addicted rats (Avena et al. 2008). More specifically, these rats exhibited "binge-like behavior," when given a palatable sugar solution; "withdrawal" symptoms when access to the sugar solution was removed; "craving" for sugar; and locomotor "cross-sensitization" from sugar to abuse of drugs. These effects seemed specific to sugar, in that providing a high-fat diet instead of sugar did not induce such addictive-like behaviors (Avena et al. 2008).

9.8   The Role of Sugar in Controlling Pain

Sugar has been known to reduce pain in human beings. Relief from pain attributed to sweet-tasting foods, and fluids have been referred to since biblical times. As an example, to lessen the pain of ritual circumcision, in Jewish tradition, a drop of sweet wine is placed in the mouth of the new born babies, while in the Muslim tradition, pieces of dates are rubbed inside the infant's mouth (Katme 1995). More recent validation of these ancient prescriptions comes from research assessing the effects of short-term intake of palatable fluids on pain sensitivity in both humans and experimental animals. The first scientific investigation of sugar's pain-relieving properties was completed by Blass et al. (1987) who found that infusions of sucrose into the oral cavity of infant rats reduced pain sensitivity and led to a diminution of distress vocalizations when these animals were isolated from their dam or siblings. Shortly thereafter, this research was extended to humans in whom it was found that small quantities of sucrose placed on the tongue of 1–3 days old infants reduced crying in response to a painful stimulus like circumcision or a heel stick. In the intervening years, these results have been repeatedly confirmed. These studies indicate that in human infants, pain relief begins approximately 30 s after exposure to a sweet solution and lasts at least 4 min after exposure (Blass and Shah 1995). An important question raised by these initial experiments was whether the pain-relieving properties of sugars are restricted to sucrose or extend to other simple carbohydrates. Results on this indicated that in both rat pups and human infants, sucrose and fructose were most effective in producing analgesia followed by glucose. In contrast, lactose, the sugar found in milk, failed to relieve pain in either rats or humans (Blass and Smith 1992). These findings are interesting when compared with those discussed earlier because they showed that human infants can discriminate between sucrose, fructose, glucose, and lactose. All taken together, these studies indicate that the palatability of the sugar involved rather than its nutritive consequences may be critical in determining analgesic responses. Although the analgesic actions of sucrose tend to wane with age, the consumption of palatable foods and fluids can reduce pain sensitivity in adults as well as in younger organisms (Kanarek and Arrington 2004). As an example, Mercer and Holder (1997) measured adults' sensitivity to pressure-induced pain and found that women who consumed a palatable sweet food displayed increased tolerance of pain relative to women consuming either an unpalatable or neutral-tasting food. However, in men, no such differences in pain sensitivity were observed as a function of nutrient palatability.

References

Avena, N. M., Rada, P., & Hoebel, B. G. (2008). Evidence for sugar addiction: Behavioral and neurochemical effects of intermittent, excessive sugar intake. *Neuroscience and Biobehavioral Reviews, 32*, 20–39.

Bellisle, F. (2004). Effects of diet on behavior and cognition in children. *British Journal of Nutrition, 92*, S227–S232.

Benton, D. (2003). Carbohydrate, memory and mood. *Nutrition Reviews, 61*, S61–S67.

Blass, E. M., & Shah, A. (1995). Pain-reducing properties of sucrose in human newborns. *Chemical Senses, 20*, 29–35.

Blass, E. M., & Smith, B. A. (1992). Differential effects of sucrose, fructose, glucose, and lactose on crying in 1–3 day-old human infants: Qualitative and quantitative considerations. *Developmental Psychology, 28*, 804–810.

Blass, E. M., Fitzgerald, E., & Kehoe, P. (1987). Interactions between sucrose, pain, and isolation distress. *Pharmacology, Biochemistry & Behavior, 26*, 483–489.

Centers for Disease Control and Prevention. (2013). Attention Deficit/Hyperactivity Disorder (ADHD). *Data & Statistics*. Available at: http://www.cdc.gov/ncbddd/adhd/data.html. Accessed 25 Mar 2015.

Christensen, L., & Pettijohn, L. (2001). Mood and carbohydrate cravings. *Appetite, 36*, 137–145.

Darby, W. J., Ghalioungui, P., & Gravetti, L. (1977). *Food: The gift of Osiris*. London: Academic.

Guo, X., Park, Y., Freedman, N. D., Sinha, R., & Hollenbeck, A. R. (2014). Sweetened beverages, coffee, and tea and depression risk among older US adults. *PLoS ONE, 9*, e94715. https://doi.org/10.1371/journal.pone.0094715.

Kanarek, R. B., & Carrington, C. (2004). Sucrose consumption enhances nicotine-induced analgesia in male and female smokers. *Psychopharmacology, 173*, 56–63.

Kanarek, R. B., Mandillo, S., & Wiatr, C. (2001). Chronic sucrose intake augments antinociception induced by injections of mu but not kappa opioid receptor agonists into the periaqueductal gray matter in male and female rats. *Brain Research, 920*, 97–105.

Kanarek, R. B., D'Anci, K. E., Mathes, W. F., Yamamoto, R., Coy, R. T., & Leibovici, M. (2005). Dietary modulation of the behavioral consequences of psychoactive drugs. In H. R. Lieberman, R. B. Kanarek, & C. Prasad (Eds.), *Nutritional neuroscience* (pp. 187–206). New York: CRC Press.

Katme, A. M. (1995). Analgesic effects of sucrose were known to the prophet. *British Medical Journal, 311*, 1169.

Lieberman, H. R. (2001). The effects of ginseng, ephedrine, and caffeine on cognitive performance, mood and energy. *Nutrition Reviews, 59*, 91–102.

Mercer, M. E., & Holder, M. D. (1997). Antinociceptive effects of palatable sweet ingesta on human responsivity to pressure pain. *Physiology & Behavior, 61,* 311–318.

Reid, M., & Hammersley, R. (1995). Effects of carbohydrate intake on subsequent food intake and mood state. *Physiology & Behavior, 58,* 421–427.

Reid, M., & Hammersley, R. (1999). The effects of sucrose and maize-oil on subsequent food intake and mood. *British Journal of Nutrition, 82,* 447–455.

Saris, W. H. M. (2001). Very-low –calorie diets and sustained weight loss. *Obesity, 9,* 295S–301S.

Vitale, M. A., Chen, D., & Kanarek, D. B. (2003). Chronic access to a sucrose solution enhances the development of conditioned place preferences for fentanyl and amphetamine in male Long-Evans rats. *Pharmacology, Biochemistry & Behavior, 74,* 529–539.

Westover, A. N., & Marangell, L. B. (2002). A cross-national relationship between sugar consumption and major depression? *Depression and Anxiety, 16,* 118–120.

Willner, P., Benton, D., Brown, E., Cheeta, S., Davies, G., Morgan, J., & Morgan, M. (1998). "Depression" increases "craving" for sweet rewards in animal and human models of depression and craving. *Psychopharmacology, 136,* 272–283.

Wolraich, M. L., Lindgren, S. D., Stumbo, P. J., Stegink, L. D., Appelbaum, M. I., & Kiritsy, M. C. (1994). Effects of diets high in sucrose or aspartame on behavior and cognitive performance of children. *New England Journal of Medicine, 330,* 301–307.

Chapter 10
The Caffeine, Methylxanthines, and Behavior Linkages

Abstract The chapter would, at first, discuss caffeinated beverages – a historical perspective – and continue with a discussion on, principally, the caffeine, methylxanthines, and behavior linkages, and a discussion on coffee and tea, in particular, would follow. Additionally, it would also discuss different energy drinks, how methylxanthines metabolize in the body, the physiological effects of caffeine on the cardiovascular system, the modus operandi of caffeine within the central nervous system, and finally, intoxication and addiction by caffeine. It would conclude with what the verdict is on coffee.

Keywords Caffeine · Methylxanthines · Behavior · Coffee · Tea · Energy drinks · Intoxication · Addiction

Starting the day, after waking up, a cup of coffee, either black or with milk and sugar, is a "must" for millions around the world. While 75% of American adults admit to drinking coffee once a week, 63% say they drink it each day (National Coffee Association 2013). The Dutch are top coffee drinkers with per capita consumption at 2.4 cups a day, while Americans are at 1 cup a day. Despite widespread consumption of coffee, tea is still the most widely consumed beverage, worldwide. While Turkey has the highest per capita consumption of tea, the UK consume an average of 3–4 cups of tea each day, which translates to over 4 lb./year per person, or eight times the amount consumed by Americans. Young people often find the taste of coffee and tea to be unpalatable. However, children and adolescents drink substantial quantities of soft drinks both at home and while being away. Coffee consumption has increased of late among youngsters due to the availability of specialty coffee.

Though coffee, tea, and soft drinks vary widely in taste and nutrient composition, they share one common factor, that is, all of them contain chemicals methylxanthines. There are a large number of methylxanthines, but only three are commonly found in foods: caffeine, theophylline, and theobromine. Caffeine is naturally found in coffee, kola nuts, tea, and chocolate and is an added ingredient in over 70% of the soft drinks manufactured in the USA (Bernstein et al. 2002). Aside from caffeine, theophylline is found in tea, and theobromine in chocolate. Guarana or Brazilian cocoa, an ingredient in many energy drinks, contains caffeine, theophylline, and

© Springer Nature Switzerland AG 2020
K. P. Nair, *Food and Human Responses*,
https://doi.org/10.1007/978-3-030-35437-4_10

theobromine. All of the three compounds produce significant physiological effects in humans. The action of these drugs on CNS is what contributes most significantly to their use. As a group, methylxanthines are the most commonly consumed psychoactive substances in the world (James 1997). They are basically strong stimulants, best avoided, if one can.

10.1   Caffeinated Beverages: A Historical Perspective

10.1.1   Coffee (Coffea arabica)

A young Ethiopian goat herder called Kaldi discovered the coffee plant in the ninth century, according to legend. One evening, Kaldi found his animals behaving in a most unusual manner – running around, butting each other, and dancing on hind legs – and darting off at the sight of the young herder. These antics lasted through the night, which were repeated the following day. Careful observations revealed that the frenetic behavior of the animals resulted from the intake of the glossy green leaves and red berries of a tree he had never seen before. A curious youth, Kaldi, tasted the plant and was rewarded with a burst of energy and feeling of euphoria. In his quest to sustain these positive effects, Kaldi became the first habitué of coffee. As word of these magical trees spread, the plant became an integral part of Ethiopian culture (Luttinger and Dicum 2006).

In Ethiopia and other parts of Africa, primarily in the Arab world, the berries of the plant were originally chewed. By the fifteenth century, the berries were roasted and mixed with boiled water to make a beverage called *qahwah* which in Arabic means coffee. In Kashmir, in northern India, the same word is used, especially among Muslim tradesmen, and the *qahwah* is a very popular drink. It is drunk, as such, with no sugar or milk added. As people got experienced with the plant, its value as a medication with the potential to alleviate several maladies, including stomach distress, fatigue, and pain, became generally recognized (Luttinger and Dicum 2006).

Until the seventeenth century, coffee remained a monopoly of the Arab world, when Europeans began to put their feet in Africa and the Middle East, though foreigners were not allowed to visit coffee farms. Extreme care was taken against coffee beans being exported out of Africa or Middle East. In the middle of the seventeenth century, Dutch spies managed to smuggle coffee plants from the Middle East. The Europeans after colonizing much of Africa got into the coffee business, and the coffee trade was practically controlled by them. In the ensuing 100 years, coffee drinking became commonplace throughout Europe and North America, with "Coffee Houses" becoming important political and cultural venues in Holland, Italy, France, Austria, and England (Luttinger and Dicum 2006). Though coffee growing was confined to the African continent, the trade monopoly shifted to Europe. Even, as late as the twenty-first century, while Africa remained the biggest coffee producer, the price was dictated from European capitals. There is a parallel here between coffee and spices. The Portuguese came to India in search of the spices at a time when spice trade was controlled by the Arabs (Nair 2013a, b). It is the climatic constraints that prevented spices like black pepper, cardamom, ginger, and turmeric remaining in India and not spreading to other parts of the world (Nair 2011, 2013a, b).

It was in Boston the first "Coffee House" was opened in 1689. Coffee intake increased with the growth of the USA, and coffee became a staple of life in both cities and the countryside. By the middle of the 1800s, enterprising entrepreneurs began roasting green coffee beans and packaging them into 1 lb. bags, further increasing the availability and popularity of the beverage. At the beginning of the last century, manufacturers developed a process for refining coffee crystals from brewed coffee which could be dissolved in hot water as "instant" coffee. Although instant coffee did not have the invigorating smell, taste, or body of coffee brewed from fresh beans, it did resemble the real coffee. During World War I, instant coffee provided the soldiers with a reasonable source of warmth and caffeine. By the middle of the century, freeze drying permitted instant coffee to more closely mimic real coffee from dried beans. In 1978, instant coffee represented 34% of all coffee consumed in the USA. However, with the rise in gourmet and specialty coffees in recent years, a decline in the popularity of instant coffee has been observed. It now accounts for only 7% of coffee intake in the USA, though it is extremely popular in the UK, where tea, of course, is king (Magazine Monitor 2014).

Nair (2010) has written extensively on coffee, its genetic resources, breeding programs, field management, pests and diseases, consumer choice and coffee adaptation, research and development, and a look into the future of coffee.

10.1.2   Tea (Camellia sinensis L.)

Like coffee, the origins of tea are also shrouded in myth. One of the most popular suggests that the founder of Zen Buddhism, Daruma, fell asleep while meditating. Upon awakening, he was so dejected that he cut off his eyelids and cast them on the ground. The first tea plant grew from this spot, and the leaves were used to brew a beverage that produced alertness (Ray and Ksir 2013). Chinese legend says that the Emperor Shen Nong accidentally discovered tea when tea leaves from twigs he used for a fire rose up on a column of hot air and landed in the water that was being boiled. Portuguese traders and priests visiting China introduced tea to Europe in the sixteenth century and to Great Britain some hundred years later. Initially expensive and taxed to boot, tea did not become an everyday beverage in Great Britain and Ireland until the late nineteenth century (Lysaght 1987).

Nair (2010) has written an extensive review on tea, covering its origin and history; different practices of tea consumption; it's botany, taxonomy, and genetics; growing conditions; nutrient requirements; nursery practices; crop management; propagation and genetic improvement of tea stocks; the evolution of rational pest and disease management; variations in tea manufacture; antioxidants in the main types of tea; the economics of tea production and global tea trade; research and development related to tea; and a look into tea's future and SWOT: the final word.

Whatever be the origin of tea, it has been consumed for millennia. In fact, for much of its history, drinking hot tea was safer than drinking water, due in part to the fact that the water to make tea needed to be boiled. Indeed, when drinking tea in the days before modern plumbing, the first "cup" of tea was poured to rinse out the cup so as to cleanse it for subsequent use.

The second largest source of caffeine is tea in the American diet. Approximately 15% of the caffeine consumed in the USA comes from tea. By weight, tea leaves contain more caffeine than an equal amount of coffee beans. However, since a smaller quantity of tea leaves is only required to brew a cup of tea than the quantity of coffee beans required to make a cup of coffee, a cup of tea generally contains less caffeine (30–50 mg) than a cup of coffee. A cup of tea also contains 1–2 mg of the methylxanthine known as theophylline.

10.1.3 The Different Energy Drinks

There are many energy drinks in the market, especially that of the USA. It is far less popular in developing countries like India, though one can see them in metros. Compared to either coffee or tea, the energy drinks are only of recent origin. These grew out of the soft drink industry. Coca-Cola derived its name from its two active ingredients, cocoa leaves and kola nuts, both of which are stimulants, and Pepsi was originally marketed as an "invigorating" drink which was expected to boost energy. Concerns about cocaine (from fresh coca leaves) as well as caffeine (from kola nuts) led to changes in their respective formulas, and the soft drinks were then marketed more for their taste than for promoting health. In fact, these soft drinks have become the most important health hazard of the present century. Children who get addicted to such drinks put on weight and get certainly unhealthy. Yet, with all possible marketing tricks, where even celebrities like popular cinema stars and sportsmen are engaged by paying them huge promotion royalties, the companies promote these soft drinks and the naïve viewers of these promotional advertisements on television screens and cinema halls fall as preys, as is beginning to happen in a country like India. The companies harvest phenomenal profits at the expense of the misguided consumers, at their health risks, as these energy drinks are nothing but short-term stimulants.

In 1949, a soft drink fortified with thiamin, niacin, and potassium, along with the requisite sugar and caffeine contents, was developed and marketed by Dr. Enuf® with a label that read "Dr Enuf......... is energy" (Sauceman 2009). In 1985, Jolt Cola® was introduced with the slogan, "All the sugar and twice the caffeine," and was targeted at the students and young professionals as a soft drink that would promote alertness.

10.1.4   Caffeine and Its Varied Sources

A glass of chocolate milk, a cup of cocoa, or a chocolate bar, all of which contain 5–10 mg caffeine and 250 mg theobromine (Harland 2000). Caffeine from chocolate makes up only a small part of total caffeine intake in adults. However, chocolate provides the major source of methylxanthines for many children. In addition to being components of coffee, tea, and energy drinks, caffeine, theobromine, and theophylline are found in a variety of foods, such as yogurt, ice cream, and energy bars, and pharmaceutical products, such as analgesics, allergy medications, and weight control products. The recent spurt in snack food consumption is another source of caffeine intake, especially among children. Though indirect, caffeine consumption through snacks can pose health hazards among children.

10.2   How Do Methylxanthines Metabolize in the Body?

There is quite a chemical similarity in the metabolism of the three methylxanthines, namely, caffeine, theophylline, and theobromine. Hence, the following discussion will only confine to caffeine, which is the most widely consumed methylxanthine. Over 99% of the caffeine is rapidly absorbed from the intestines of adults when it is consumed as a beverage. Within a span of just 5 min, caffeine is distributed to all parts of the body including the brain. Additionally, caffeine easily crosses the placental barrier to enter fetal tissue and passes from mother to infant in breast milk (Blanchard et al. 1992). Once consumed, peak effects are seen within 15 min to 2 h. The wide variation in peak effects is related to speed of gastric juice emptying and the presence of other dietary components, such as fiber in the digestive tract (Mandel 2002). Caffeine is metabolized by the liver into 1-methyluric acid and 1-methylxanthene. The kidneys, primarily, excrete these metabolites (Arnaud 2011). The "half-life" of caffeine is the time it takes to eliminate one-half of the consumed caffeine varieties among individuals but is normally 3–7 h in the healthy ones. However, a number of factors, including reproductive status, age, disease, and cigarette smoking, can influence metabolism. For instance, the half-life of caffeine increases to 6–14 h in women who take oral contraceptives or who are in the last trimester of their pregnancy. In infants it is approximately 50 h, and in individuals with liver disease, it is approximately 60 h than in healthy adults. By contrast cigarette smoking reduces the half-life to approximately 3 h.

10.3   Physiological Effects of Caffeine

10.3.1   Cardiovascular System

The potency of caffeine, theophylline, and theobromine has similar physiological effects throughout the human body. However, the potency of these drugs varies according to the physiology of the system under consideration as shown in Table 10.1.

The actions of caffeine and other methylxanthines on the cardiovascular system are complex and quite often antagonistic. A primary reason for the difficulty to precisely understand this is that the effect of the drug depends on important factors concerning the individual, such as his/her age, the history of the drug consumption, the dose involved, and the drug administrative route. Clinical studies point out to an increase in the individual's blood pressure and heartbeat following caffeine intake (Noordzij et al. 2005). The effects of blood pressure are most often observed at doses of 250 mg of caffeine or more. Although the pressor effect of caffeine is seen in both men and women, irrespective of age, race, or the individual's blood pressure, the effect is most pronounced in the elderly, hypertensive, and caffeine-naïve individuals (James 2004).

Though increase in blood pressure follows caffeine intake, epidemiological investigations have only yielded contradictory results with regard to caffeine consumption and hypertension. These inconsistent findings reflect the difficulties encountered with accurate determination of caffeine intake and development of tolerance to the drug's pressor action, discussed earlier, and the possibility of confounding factors, such as smoking and body weight, in moderating blood pressure. Controversy also surrounds the contribution of caffeine intake to high cholesterol levels and the development of cardiac arrhythmias. While concern continues about

Table 10.1 Comparative potency of methylxanthines on human physiology

Details	Caffeine	Theophylline	Theobromine
CNS stimulation	1	2	3
Respiratory stimulation	1	2	3
Diuresis	3	1	2
Cardiac stimulation	3	1	2
Smooth muscle relaxation	3	1	2
Skeletal muscle relaxation	1	2	3

Note: The number 1 denotes the most potent effect

the contribution of caffeine intake to the development of heart disease, currently available research data suggest that moderate caffeine intake (about 400 mg/day) does not have adverse effects on cardiovascular health (Nawrot et al. 2003).

Some important effects of caffeine on human physiology:

Smooth muscle: Methylxanthines relax many smooth muscles in the human body, including those found in the bronchi of the lungs. Since theophylline has the capacity to dilate the bronchi, it is used in the treatment of asthma. The drug is prescribed as a prophylactic therapy for asthma and also used as an adjunct in the treatment of prolonged asthma attacks (Spina 2003). In addition, as a result of the ability of theophylline and related compounds to stimulate the respiratory system, both theophylline and caffeine have been widely used to prevent episodes of the loss of effective breathing (sleep apnea) in preterm infants (von Poblotzki et al. 2003).

Gastrointestinal system: Caffeine has been found to stimulate secretion of gastric acid and pepsin, the enzyme which triggers the breakdown of proteins in the stomach. As a result, coffee intake is often considered detrimental to individuals suffering from gastric ulcers (Marotta and Floch 1991). However, as both caffeinated and decaffeinated coffee increase gastric secretions, it appears that caffeine alone is not the culprit and there are additional components in coffee which contribute to the beverages actions on the gastrointestinal system.

Renal system: It has long been recognized that coffee, like other methylxanthines, are diuretics. Acute ingestion of 250–300 mg caffeine results in the short-term stimulation of urine output and Na excretion in individuals deprived of caffeine for days or weeks on end. Regular intake of caffeine, however, is associated with the development of tolerance of the diuretic effects. Hence, the action of these on the renal system is reduced in individuals, who regularly consume coffee or tea (Maughan and Griffin 2003).

Caffeine and reproduction: Many have cautioned the deleterious effects of caffeine in pregnancy. It can induce infertility, miscarriage, low birth weight of babies, and birth defects (Schardt and Schmidt 1996). The most frequently observed abnormalities are facial and limb defects. However, relatively high intake of caffeine was required to produce these effects. Lower doses of caffeine had negligible effects on renal development, and these effects were seen in laboratory animals, where caffeine served as a teratogen.

Caffeine and physical performance: Athletes and sportsmen and women are the ones who most benefit from caffeine intake. Additionally, caffeine has proven to be an effective strategy to maintain physical performance in military and related situations where sustained operations are necessary. Caffeine increases heartbeat rate, respiration, blood pressure, and the rate of blood glucose absorption, which together contribute to the positive effects of the drug on physical performance. In addition, following coffee intake, energy derived from fat is increased, while energy derived from carbohydrates is decreased, allowing individuals to sustain physical activity for a longer time. Another positive aspect of coffee intake is that it may reduce the pain from strenuous physical activity. Caffeine's pain-relieving properties stem, at least in part, from the ability of the drug to stimulate the release of beta-endorphin – the body's natural "painkiller" (Laurent et al. 2000). The positive effects of caffeine on physical activity are most apparent in, but not limited to, endurance activities. Caffeine at doses ranging from 2 mg to 9 mg/kg of body weight improves performance in a variety of activities, like running cross-country races, skiing, cycling, etc., lasting as short as 60 seconds to 2 h in both trained and untrained athletes. While caffeine can boost performance of endurance activities, it has but minimal effect in enhancing maximal abilities on tasks which require high power output, like lifting and carrying, although caffeine consumption may prolong the time until fatigue sets in (Bell and McLellan 2003).

Coffee and sleep: For many, a cup of coffee with dinner can lead to a night of scanty sleep. Caffeine delays sleep onset, shortens sleep time, reduces the average depth of sleep, and worsen the subjective quality of sleep. At high enough doses, caffeine can lead to insomnia. However, while for many caffeine may distort sleep pattern, the effect is not universal. Generally, those who regularly consume caffeine have fewer problems falling asleep after an evening cup of coffee or tea than those who abstain from caffeine. Moreover, the effect of caffeine on sleep is dose-dependent with larger quantities of caffeine which have more negative effects on sleep than smaller quantities. Although caffeine can alter patterns of sleep, there is no strong evidence which indicates that alterations in sleep patterns have a significant effect on the daytime behavior of the individual concerned (Roehrs and Roth 2008). Caffeine-induced sleep disturbances while not posing much of a problem among adults can be a source of disturbance in children. Pollak and Bright examined the relationship between caffeine consumption among seventh, eighth, and ninth graders and sleep disturbances. On average, respondents consumed at least one caffeine-containing item a day, contributing a little over half of the caffeine consumed. Boys consumed more caffeine than girls. Caffeine intake was associated with shorter nighttime sleep, more day time naps, and more interrupted sleep patterns. While only a single study survey, this report is an important step in quantifying the actual patterns of caffeine intake in children and suggests that caffeine consumption may disrupt sleep patterns in children.

10.4   Methylxanthines and Their Neurophysiological and Human Behavioral Reactions

The action of methylxanthines on the CNS on human behavioral aspects is the most critical, though their actions on human physiological aspects are important. Among caffeine, theophylline, and theobromine, it is caffeine that has the most potent effects on CNS. A negligible amount of caffeine consumption, as negligible as

150 mg in a 12 ounce mug of coffee, stimulates the cortex activity, the area of the brain involved in human higher mental function. Higher dose (500 mg) can stimulate the medulla, a portion of the hindbrain important in controlling respiration, cardiovascular function, and muscular activity, besides the cortex. When intake reaches 1 g, restlessness and insomnia can result, which may be accompanied by sensory disturbances including ringing sensation in the ears and flashes of light. More than a century of research has confirmed the belief that caffeine has the power to maintain humans' alertness and improve performance on a variety of mental tasks (Lieberman 2003). But, there are contradictory reports in scientific literature on this score. It might be educative to examine why this is the case. The following are some of the reasons:

First, most studies investigating the behavioral consequences of caffeine intake have used only a single dose of the drug. Too low a dose may not be sufficient to produce a behavioral response, while too high a dose might lead to side effects. In general, data point to an inverted U-shaped dose-response curve for caffeine's effect on human behavior. Doses ranging from 100 to 600 mg generally increase alertness and concentration, while still higher doses might lead to restlessness, feelings of anxiety, and deterioration in performance of cognitive tasks. Other differences in participant populations among investigations may also contribute to whether caffeine does or does not improve cognitive performance. Of particular importance is the amount of sleep individuals have had being tested for the effects of caffeine on behavior. Age, sex, and personality characteristics also may play a role in determining responses. For example, elderly individuals are more sensitive to caffeine's behavioral actions than young and middle-aged adults (Johnson-Kozlow et al. 2002). Given the widely held view that caffeine has stimulant effects, there is the possibility that expectancy or placebo effects can confound the results of a study. In many investigations, caffeine has been provided in the form of coffee or tea with decaffeinated coffee or tea used as the control treatment or placebo. If these studies fail to find an effect of caffeine on behavior, it cannot be determined if this is because caffeine has no effect or that a placebo response to the decaffeinated beverage prohibited the drug's action from being observed (Smith 2005). To overcome the possible confounds of expectancy effects when assessing the role of caffeine on behavior, studies have either supplied the drug as a capsule or pill or used a beverage not normally associated with the drug, such as a fruit-flavored drink (Rogers et al. 2003).

The debate is on whether the effect of caffeine on behavior originates from the drug *per se* or it is a reflection of the reversal of the negative consequences of caffeine withdrawal. With regular caffeine users, withdrawal of caffeine intake is associated with headache, irritability, mental confusion, and general fatigue. These symptoms generally begin 12–24 h after the last cup of coffee or tea is consumed. Although contradictory data exist, the overwhelming majority of research investigations support the conclusion that intake of moderate levels of caffeine has beneficial effects on aspects of cognitive behavior. In general, caffeine results in an increase in alertness and a decrease in fatigue. However, when the effects of caffeine are examined in more detail, it appears the drug preferentially facilitates performance of tasks requiring sustained attention such as simple reaction time and vigilance tasks (Santos et al. 2014). There is but scanty evidence that caffeine intake enhances intellectual capabilities, except, perhaps, the fact that it boosts normal performance levels when they are lowered by fatigue of the body. This is, in particular, applicable to sleep-deprived individuals. The best way to reverse the adverse effects of sleep deprivation is simply to go to sleep; even a short nap of 15–30 min can alleviate some of the detrimental consequences of sleep deprivation. However, the longer the sleep episode, the greater is the restoration in cognitive function. If sleep is not an option, consuming a cup of coffee or tea or taking even a caffeine pill can help lessen some of the impairments in cognitive performance associated with sleep deprivation. Caffeine can also reverse the impairments in cognitive function associated with minor illnesses, such as cold, influenza, or fatigue related to shifts in work schedule, like in jet lag, with pilots and air hostesses or cabin crew (Lorist and Tops 2003).

10.5   Caffeine: Its Modus Operandi Within the Central Nervous System

The biological reaction in the human body that caffeine intake triggers is through several routes – altered cellular Ca conduction, enhanced cyclic adenosine monophosphate (AMP), and antagonism of adenosine receptors. The first two reactions will occur when caffeine is consumed at rates higher than ordinarily consumed by most individuals and, thus, are not of major importance in determining the caffeine intake-behavior interrelationship. The scientific consensus is that the majority of caffeine-related behavioral changes are related to the interactions of the drug with adenosine receptors. Adenosine is formed during the breakdown of adenosine triphosphate (ATP), the primary energy source for most cells in the body, and is found throughout the nervous system and is considered to be a neuromodulator, which produces its behavioral effects by inhibiting the conduction of messages at synapses which use other neurotransmitters, such as dopamine and norepinephrine. Receptors for adenosine are present in the gastrointestinal tract, heart, blood vessels, respiratory system, and brain. Stimulation of peripheral adenosine receptors decreases intestinal peristalsis, blood pressure, and heartbeat rate and increases bronchial tone. In the brain, adenosine inhibits neuronal activity which results in feelings of fatigue and behavioral depression. These effects which are opposites of many of caffeine's known actions have led to the hypothesis that much of caffeine's effects can be attributed to the ability of the drug to act as an antagonist at adenosine receptors (Fisone et al. 2004). In support of this hypothesis, when consumed at levels comparable to those found in 1–3 cups of coffee, caffeine occupies 50% of the adenosine receptors in the brain. By blocking receptor sites, caffeine prevents adenosine from inhibiting neuronal firing, thereby allowing increased stimulation of neuronal activity and behavioral responses (Davis et al. 2003). Alterations in adenosine receptors may also play a role in the development of tolerance to and physical dependence on caffeine, which occurs following prolonged use of the drug (see Table 10.2). Chronic caffeine consumption is accompanied by an increase in the adenosine receptors (Boulenger et al. 1983). As a result of this, a new balance between endogenous adenosine and the presence of exogenous caffeine occurs leading to a reduction in some of the physiological and behavioral actions of the drug tolerance. If this balance is altered by severely decreasing or abruptly stopping caffeine use, then the excess adenosine receptors would no longer be blocked by caffeine, and the physiological response to adenosine would be exaggerated resulting in symptoms of caffeine withdrawal.

10.6 Intoxication and Addiction by Caffeine

Moderate amounts of caffeine intake while producing only benign reactions in humans, excessive intake will lead to negative consequences. Excessively high intake, as much as 10 g, can lead to vomiting and convulsions and, in some cases, can even lead to death. Additionally, regular intake of small amount of caffeine 1 g/day can lead to nervousness, irritability, loss of appetite, neuromuscular tremors, and vomiting. Repeated high intake of caffeine in excess of 1 g/day can lead to a constellation of symptoms similar to anxiety neurosis. This constellation of symptoms, listed in Table 10.2, has been termed "caffeine intoxication" and is classified as an organic mental disorder with known cause (caffeine consumption) in the DSM-5 (*Diagnostic and Statistical Manual of Mental Disorders* 2013). Caffeine is a drug which produces physiological, psychological, and behavioral effects in the human body, but does that make it a drug of abuse? There has been considerable debate questioning the abuse potential of caffeine. Many people feel that they are dependent on caffeine – everyone knows at least one individual who cannot face the day without a cup of coffee in the morning. However, addiction and dependence are terms which are usually used as interchangeable ones which describe behaviors relating to drugs, such as heroin, nicotine, alcohol, and cocaine.

Table 10.2 Diagnostic criteria for caffeine intoxication

Recent consumption of caffeine, in excess of 250 mg (more than 2–3 cups of brewed coffee)
Five or more of the following symptoms developing during or shortly after caffeine use
a. Restlessness b. Nervousness c. Excitement d. Insomnia e. Flushed face f. Diuresis g. Gastrointestinal disturbances h. Muscle twitching i. Rambling flow of thought and speech j. Tachycardia or cardiac arrhythmia k. Periods of inexhaustibility l. Psychomotor agitation
The symptoms in criterion number 2 cause significant distress or impairment in social, occupational, or other important areas of functioning
The symptoms are not due to a general medical condition and are not better accounted for by another mental disorder such as intoxication with another substance

Source: Diagnostic and Statistical Manual of Mental Disorders 2013

It is important to understand the implications and meanings of some of the terms used in psychopharmacology, before one can discuss how a substance is classified as addictive vis-à-vis caffeine as one. According to psychopharmacologists and clinicians, for a drug to be classified as addictive, it *must* meet a set of primary criteria and *may* meet a set of secondary criteria. For instance, one hallmark of drug addiction is escalating use of a drug and loss of control over drug taking behavior. Typically, people who are addicted to a drug need more and more of the drug to achieve the same effect – a phenomenon called tolerance. Moreover, although originally taking the drug may have produced pleasurable or reinforcing effects, addicts maintain drug-taking behavior to avoid feeling ill – withdrawal symptoms – when ceasing to take the drug. Thus, drug addiction is indicative of a cycle of behaviors which include pursuit of pleasurable effects and avoidance of negative effects and often to the exclusion of other pursuits and to the overall detriment of the individual.

The first and foremost point to be taken into consideration while judging whether a substance is a confirmed addictive is whether or not it produces a pleasurable or reinforcing effect on the individual who uses it. Caffeine's reinforcing properties are presumed to be related to the drug's ability to produce physiological arousal, including enhanced alertness, mood, and concentration. Some scientists have argued that the reinforcing effect of caffeine is similar to characteristics, not in magnitude though, to psychostimulant drugs such as cocaine or amphetamine. Others (Daly and Fredholm 1998) have argued that while the effect of caffeine is stimulatory, the reinforcing effects are rather minimal. More importantly, unlike typical drug abuse, individuals do not normally need to consume increasing amounts of caffeine (drug abuse) but rather use the drug at consistent and moderate levels (drug use). The issue of withdrawal may be the most important consideration in discussing the addictive potential of caffeine. The subjective signs of caffeine withdrawal are in opposition to the effects of taking caffeine, which include headache, fatigue, depression, difficulty to concentrate, irritability, and sleepiness. While some do not suffer the adverse consequences of abstaining from caffeine, 40–70% of the individuals who are addicted to it who tried to quit effective use experience symptoms of caffeine withdrawal. These symptoms normally begin within 12–14 h after termination of caffeine intake (Juliano and Griffiths 2004). Peak withdrawal symptoms are reported normally 24–48 h after abstinence. Fortunately for those trying to abstain from caffeine, symptoms of withdrawal normally are relatively mild and subside, within a few days. However in some, withdrawal symptoms can lead to impairments in carrying out daily functions smoothly and may continue for weeks or sometimes even for months. Withdrawal symptoms may lead some to return to regular use and contribute to the potentially addictive qualities of the drug.

10.7   What Is the Verdict on Coffee?

In conclusion, one cannot pass the judgment that caffeine is an addictive substance (drug) like cocaine or heroin, and it will be foolhardy to expect that millions over the world will give up using coffee, if a perception is developed that it is an addictive "drug" subject to misuse and/or abuse. Perhaps the best advice one could proffer is that used in moderation, coffee could be one of the "simple pleasures of life."

References

Arnaud, M. J. (2011). Pharmacokinetics and metabolism of natural methylxanthines in animal and man. *Handbook of Experimental Pharmacology, 200*, 33–91.

Bell, D. G., & McLellan, T. M. (2003). Effect of repeated caffeine ingestion on repeated exhaustive exercise endurance. *Medicine & Science. Sports & Exercise, 35*, 1348–1354.

Bernstein, G. A., Carroll, M. E., Thuras, P. D., Cosgrove, K. P., & Roth, M. E. (2002). Caffeine dependence in teenagers. *Drug and Alcohol Dependence, 66*, 1–6.

Blanchard, J., Weber, C. W., & Shearer, L. E. (1992). Methylxanthine levels in breast milk of lactating women of different ethnic and socioeconomic classes. *Biopharmaceutics & Drug Disposition, 13*, 187–196.

Boulenger, J. P., Patel, J., Post, R. M., Parema, A. M., & Marangos, P. J. (1983). Chronic caffeine consumption increases the number of brain adenosine receptors. *Life Sciences, 32*, 1135–1142.

Daly, J. W., & Fredholm, B. B. (1998). Caffeine- an atypical drug of dependence. *Drug And Alcohol dependence, 51*, 199–206.

Davis, J. M., Zhao, Z. W., Stock, H. S., Mehl, K. A., Buggy, J., & Hand, G. A. (2003). Central nervous system effects of caffeine and adenosine on fatigue. *American Journal of Physiology: Regulatory, Integrative and Comparative Physiology, 284*, R399–R404.

Diagnostic and Statistical Manual of Mental Disorders. 5th edn (DSM-5). (2013). American Psychiatric Association, Washington, DC.

Fisone, G., Borgkvist, A., & Usiello, A. (2004). Caffeine as a psychomotor stimulant: Mechanism of action. *Cellular and Molecular Life Sciences, 61*, 857–872.

Harland, B. F. (2000). Caffeine and nutrition. *Nutrition, 1*, 522–526.

James, J. E. (1997). *Understanding caffeine: A biobehavioral analysis*. Thousand Oaks: Sage.

James, J. E. (2004). Critical review of dietary caffeine and blood pressure: A relationship that should be taken more seriously. *Psychosomatic Medicine, 66*, 63–71.

Johnson-Kozlow, M., Kritz-Silverman, D., Barrett-Connor, E., & Morton, D. (2002). Coffee consumption and cognitive function among older adults. *American Journal of Epidemiology, 156*, 842–850.

Juliano, L. M., & Griffiths, R. R. (2004). A critical review of caffeine withdrawal: Empirical validation of symptoms and signs, incidence, severity, and associated features. *Psychopharmacology, 176*, 1–29.

Laurent, D., Schneider, K. E., Prusacyck, W. K., Franklin, C., & Vogel, S. M. (2000). Effect of caffeine on muscle glycogen utilization and the neuroendocrine axis during exercise. *Journal of Clinical Endocrinology and Metabolism, 85*, 2170–2175.

Lieberman, H. (2003). Nutrition, brain function and cognitive performance. *Appetite, 40*, 245–254.

Lorist, M. M., & Tops, M. (2003). Caffeine, fatigue, and cognition. *Brain and Cognition, 53*, 82–94.

Luttinger, N., & Dicum, G. (2006). *The coffee book: Anatomy of an industry from crop to the last drop*. New York: The New Press.

Lysaght, P. (1987). "When i makes tea, i makes tea": The case of tea in Ireland. *Ulster Folklife,*
33, 48–49.
Magazine Monitor. (2014). *Why do Britons drink so much instant coffee?* Available at: http://www.
bbc.com/news/blogs-magazine-monitor-26869244. Accessed 25 Mar 2015.
Mandel, H. G. (2002). Update on caffeine consumption, disposition, and action. *Food and*
Chemical Toxicology, 40, 1231–1234.
Marotta, R. B., & Floch, M. H. (1991). Diet and nutrition in ulcer disease. *The Medical Clinics of*
North America, 75, 967–979.
Maughan, R. J., & Griffin, J. (2003). Caffeine ingestion and fluid balance: A review. *Journal of*
Human Nutrition and Dietetics, 16, 141–420.
Nair, K. P. P. (2010). *The agronomy and economy of important tree crops of the developing world.*
London/Burlington: Elsevier.
Nair, K. P. P. (2011). *Agronomy and economy of black pepper and cardamom.* London/Waltham:
Elsevier.
Nair, K. P. P. (2013a). *The agronomy and economy of turmeric and ginger.* London/Waltham:
Elsevier.
Nair, K. P. P. (2013b). The buffer power concept and its relevance in African and Asian soils.
Advances in Agronomy, 121, 447–516.
National Coffee Association. (2013). *2013 national coffee drinking trends.* Available at: http://
www.ncausa.Org/i4a/pages/index.cfm?page1D=924. Accessed 25 Mar 2015.
Nawrot, P., Jordon, S., Eastwood, J., Rotstein, J., Hugenholtz, A., & Feeley, M. (2003). Effects of
caffeine on human health. *Food Additives and Contaminants, 20*, 1–30.
Noordzij, M., Uiterwaal, C., Arends, L. R., Kok, F. J., Grobbe, D. E., & Geleijnse, J. M. (2005).
Blood pressure response to chronic intake of coffee and caffeine: a meta-analysis of random-
ized controlled trials. *Journal of Hypertension, 23*, 921–928.
Ray, O., & Ksir, C. (2013). *Drugs, society and human behavior* (15th ed.). New York: McGraw
Hill.
Roehrs, T., & Roth, T. (2008). Caffeine: Sleep and daytime sleepiness. *Sleep Medicine Reviews,*
12, 153–162.
Rogers, P. J., Martin, J., Smith, C., Heatheley, S. V., & Smith, H. J. (2003). Absence of reinforcing,
mood and psychomotor performance effects of caffeine in habitual non-consumers of caffeine.
Psychopharmacology, 167, 545–562.
Santos, V. G. F., Santos, V. R. F., Felippe, L. L., Almeida, J. W., & Bertuzzi, R. (2014). Caffeine
reduces reaction time and improves performance in simulated –contest of Taekwondo.
Nutrients, 6, 637–649.
Sauceman, F. W. (2009). Dr *Enuf*: A new age neutraceutical with a patent medicine pedigree. In
The place setting (pp. 89–97). Macon: Mercer University Press.
Schardt, D., & Schmidt, S. (1996). Caffeine: The inside scoop. *Nutrition Action Newsletter, 23*,
1–7.
Smith, A. (2005). Caffeine. In H. Lieberman, R. Kanarek, & C. Prasad (Eds.), *Nutritional neuro-*
science (pp. 341–361). New York: CRC Press.
Spina, D. (2003). Theophylline and PDE4 inhibitors in asthma. *Current Opinion in Pulmonary*
Medicine, 9, 57–64.
von Poblotzki, M., Rieger-Fackeldey, E., & Schulze, A. (2003). Effects of theophylline on the patterns of spontaneous breathing in preterm infants less than 1000g of birth rate. *Early Human*
development, 72, 47–55.

Chapter 11
Alcohol, Brain Function, and Behavioral Impact

Abstract The chapter would principally discuss effect of alcohol on brain function and behavioral impact, the interrelationship between alcohol and nutrition, effect of acute alcohol intake, chronic alcohol use, alcohol influenced depression, and inter-relationship between nutrition and alcohol-induced neurotoxicity. Additionally, the chapter would also discuss the effect of alcohol on fetal development, fetal alcohol syndrome (FAS), and its behavioral consequences and would conclude with the question on drinking alcohol – "How much is too much?"

Keywords Alcohol · Brain function · Behavior · Nutrition · Acute alcohol intake · Chronic alcohol use · Fetal development · Fetal alcohol syndrome (FAS)

A central ingredient in human diet is alcohol, though it cannot be considered universal. There are millions who totally abstain from alcohol, for whatever reason one chooses to. It could be purely medical, or others like spiritual, especially for those in India with a *Hindu* background. There are interesting facts concerning alcohol. Recent chemical analyses of organics absorbed into pottery from ancient China have revealed that a mixed fermented beverage of grapes, rice, and honey was being produced as early as the 7th millennium BCE. As early as 4000 BCE, barley was fermented into beer (Cantrell 2000), and grape-bearing vines were cultivated to produce wine (Newman 2000), as shown by archaeological evidence from the Middle East. Drinking wine was a ceremonial custom in biblical times, the drink being appreciated for its intoxicating effect by diverse members of the society. Both ancient Romans and Greeks further developed viticulture (the technique of growing grapes), and wine became an integral part of their culture. In the Mughal tradition, drinking wine was a normal practice from aristocracy down to the commoner, and Omar Khayyam, the great Persian mathematician, philosopher, thinker, and poet, who lived in the Islamic Golden Age in Nishapur, Persia (then, now Iran), in the eleventh century, has written many verses in praise of wine. True distillation did not occur until the twelfth century, with large-scale production of distilled beverages, such as brandy, gin, vodka, rum, and whisky, although the ancient Greeks and Romans tried their hands at distilling alcohol. Distillation increases the alcohol content of a liquid, already containing alcohol, by heating the liquid to a point sufficient to boil alcohol, but, not water. The resulting vaporized alcohol (steam) is collected and condensed into a liquid with a much higher proportion of alcohol than before. Repeating the process can further increase the alcohol content of the resulting liquid. Though the instrumentation technology involved in the distillation process has vastly improved over the years, the original distillation assembly was the result of the genius Justus von Liebig, a great German chemist, in whose memory the Justus von Liebig University in Giessen, The Federal Republic of Germany, was established, where this author was privileged to work as a Senior Fellow of the world renowned Alexander von Humboldt Foundation during the early 1980s.

Prohibition of alcohol consumption by individuals or as a part of religious doctrine, as in Islam, has been a state policy (Volstead Act, USA) or in many states of the Republic of India, because of its detrimental effects on human health and its fallout on societal peace. Many a crime have had their origin in inebriation. This has been the experience throughout the ages. Obviously, this is what led to prohibition, but prohibition has not led to complete abstinence in alcohol consumption per se, because bootlegging prospered clandestinely. This has been the experience of the vast Indian state, as well. While not the focus of this review, there does exist evidence that moderate alcohol intake *may* reduce the risk of cardiac disease and stroke (Gronbaek 2004). In the USA, nearly 90% of adult population has consumed alcohol at one time or another of their lives, and in 2012, approximately 56% of adult Americans indicated that they consumed alcohol in the previous month (National Institute on Alcohol Abuse and Alcoholism 2014). While moderate alcohol intake is the rule for most Americans, approximately 7% of the population are heavy alcohol consumers (two or more drinks a day for men and one or more drink a day for women). Alcohol use is heaviest among young men between the ages 18 and 25. Less women than men in this age group report heavy consumption of alcohol. However, during the last decade, the number of men drinking large amounts of alcohol has remained relatively constant, while the number of young women drinking immoderately has increased substantially. Binge drinking is quite popular among men and women in this age group. Overall, it is estimated that 17 million Americans suffer from an alcohol use disorder and that alcohol-related problems account for an economic burden of approximately US $ 224 million each year, due to health care, property damage, and lost productivity (National Institute on Alcohol Abuse and Alcoholism 2014). The American Psychiatric Association includes alcohol use disorder within its section on substance-induced disorders (see Table 11.1). Depending on the number of symptoms present, the disorder is classified as mild (2–3 symptoms), moderate (4–5 symptoms), or severe (6 or more symptoms). Alcohol abuse is the third leading preventable cause of death in the USA with approximately 85000 people dying each year as a result of excess consumption of alcohol (Mokdad et al. 2004).

Table 11.1 Criteria for alcohol disorder

A problematic pattern of alcohol use leading to significant impairment, as manifested by at least two of the following disorders, occurring within a 12-month period:

Alcohol is frequently taken in larger amounts or over a longer period than was intended

There is persistent desire or unsuccessful efforts to cut down or control alcohol use

A great deal of time is spent in obtaining alcohol, using it, or recovering from its effects

Craving or a strong desire or urge to use alcohol

Recurrent alcohol use resulting in a failure to fulfill major role obligations at work or at home

Continued alcohol use despite persistent or recurrent social or interpersonal problems caused or exacerbated by the effects of alcohol

Important social, occupational, or recreational activities are given up or reduced because of alcohol use

Recurrent alcohol use in situations in which it is physically hazardous

Alcohol use is continued despite the knowledge of a persistent or recurrent physical or psychological problem which is likely to have been caused or exacerbated by alcohol use

Tolerance as defined by either of the following:

A need for markedly increased amounts of alcohol to achieve intoxication

A markedly diminished effect with continued use of the same amount of alcohol

Withdrawal as manifested by either of the following:

The classic withdrawal syndrome for alcohol, for example, sweating, hand tremor, agitation, and hallucinations

Alcohol or a related substance like benzodiazepine is taken to relieve or avoid withdrawal symptoms

Source: Diagnostic and Statistical Manual of Mental Disorders (2013)

11.1   The Interrelationship Between Alcohol and Nutrition

It would seem curious if one were to consider the importance of alcohol-nutrition interaction, as alcohol is most frequently used and thought of as a drug. However, alcohol has significant nutritional consequences and can profoundly alter the intake, absorption, digestion, and metabolism of a number of essential nutrients, which in turn, can affect brain function and behavior. To the body, alcohol provides energy. As a dietary macronutrient, alcohol is metabolized to produce 7.1 kilocalories/g consumed. In addition, some alcoholic beverages contain significant amounts of carbohydrate. Calories from alcoholic beverages typically account for 4.3% of total energy intake in men who drink moderately, with the figure at 2.4% for women (Suter 2012).

A most frequently asked question is, when alcohol is consumed, will it lead to the reduction in the intake of other nutrients to compensate for the calories obtained from alcohol? Research aimed at answering this question has led to the conclusion that calories consumed as alcohol do not lead to compensatory reductions in energy intake. Indeed, researchers typically report that food intake is greater after alcohol is imbibed than after consumption of an appropriate control beverage. Thus, it has been suggested that, unlike other macronutrients, alcohol fails to generate feelings of satiety and actually stimulates calorie intake. Both metabolic factors, primarily, and psychological, secondarily, contribute to the stimulatory effects of alcohol on food intake. Metabolically, alcohol suppresses fat and carbohydrate oxidation. This suppression indicates to the body that there is insufficient energy available, which is associated with sensations of hunger. On the psychological side, the ability of alcohol to disinhibit behavior may contribute to the increase in feeding behavior observed after alcohol consumption. In addition, alcohol consumed with a meal can enhance the pleasurable aspects of feeding behavior (Caton et al. 2004).

The logical extension from the previous data is does the short-term stimulatory effect of alcohol have an impact on long-term effect on body weight? Notwithstanding contradictory results, the consensus is that long-term moderate intake of alcohol is positively related to body weight. The consequences of alcohol on food intake are more detrimental on heavy drinkers. In chronic alcoholics, calories from alcoholic beverages can account for almost 50% of total energy intake (Suter 2012). As a consequence, chronic alcoholism is often accompanied by decreases in intake of essential nutrients, nutrient deficiencies, and reductions in body weight (Wannamethee and Shaper 2003). Chronic alcohol consumption compromises nutritional status by interfering with the digestion and absorption of essential nutrients (Suter 2012). In heavy drinkers, alcohol inhibits the production of saliva, leading to reductions in the secretion of salivary enzymes, inflammation of the tongue and mouth, and increased risk for dental and periodontal disease. As digestion progresses, alcohol slows down peristalsis in the esophagus and can damage the esophageal sphincter, leading to esophageal reflux and severe heartburn. In the stomach and intestines, alcohol damages the mucosal lining, which consequently impairs the absorption of a number of vitamins, including vitamin B_1 (thiamin), folic acid, and vitamin B_{12} (cobalamin).

Chronic alcoholism affects other digestive organs including the pancreas and liver, the latter being the most affected. Heavy drinkers invariably end up with liver cirrhosis. It can also lead to inflammation of the pancreas, a disease known as pancreatitis, resulting in a reduction in the secretion of the pancreatic enzymes, contributing to nutrient malabsorption. Liver cirrhosis can interfere with the metabolism of a number of essential nutrients including folic acid, pyridoxine; vitamins A, C, and D; and Na, K, and Mg (O'Shea et al. 2010). In addition, the enzymes which breakdown fat also metabolize alcohol, so that when alcohol is available, the liver preferentially uses it for energy and stores fat until all of the alcohol is used up. Over time, the infiltration of the liver by fat results in the replacement of healthy liver cells with scar tissue (cirrhosis), which ultimately reduces the amount of blood and oxygen reaching the liver and impairs the ability of the liver to perform its biochemical functions.

11.2   The Interrelationship Between Alcohol Consumption, Brain Function, and Human Behavior

11.2.1   Effect of Acute Alcohol Intake

Alcohol is quickly absorbed from the stomach and intestine into the blood and is then distributed throughout the total water in the body. The rate of alcohol absorption into the blood depends on the amount, concentration, and type of alcohol consumed. In general, the higher the alcohol concentration in the beverage, the more rapid its absorption. Diluting the alcoholic beverage will slow down absorption. However, drinking carbonated beverages, for example, champagne or whisky with soda, increases the rate of absorption because carbonation facilitates the passage of alcohol across the intestine. Food in the stomach and drugs which decrease gastrointestinal mobility or blood flow, delay alcohol absorption.

The effects of alcohol on the body are directly proportional to the concentration of alcohol in the blood, or "blood alcohol level," a parallel of which is "blood glucose level" in the case of diabetics. Blood alcohol levels are expressed in milligrams of alcohol per 100 milliliter of blood. A number of factors including sex, age, and the type of beverage consumed influence blood alcohol levels. For instance, because women generally weigh less and have less body water than men, in which alcohol can dissolve, when men and women consume identical amounts of alcohol, blood alcohol levels are usually higher in women than in men (Thomasson 1995). With regard to age, the proportion of body water decreases with aging, and hence, after consumption of similar amounts of alcohol, blood alcohol levels typically are higher in older than younger individuals. This is similar to the "concentration effect" as seen in plant cells (Nair 1996).

As alcohol is a water-soluble substance, it can readily diffuse from the blood across cell membranes into all tissues in the body. As alcohol can cross the blood-brain barrier, the concentration of alcohol in the brain parallels blood alcohol levels. Thus, alcohol follows the principle of diffusion. Alcohol affects CNS more markedly than any other system in the body. It acts primarily as a CNS depressant, and it suppresses the activity of the brain by altering properties of the neural membranes, cell metabolism, conduction of electrical impulses down the axon, and the release of neurotransmitters.

It is the most vital part of CNS, primarily involved in integrative functions, that alcohol first affects. The brain stem reticular activating system, which is important to maintain alertness, and parts of the cortex, which are particularly susceptible to the effects of alcohol. Alcohol initially depresses the activity of the inhibitory centers in the brain. Thus, at low doses, a state of disinhibition or mild euphoria is a common consequence of alcohol consumption. At higher doses, alcohol suppresses the activity of the cerebellum, resulting in slurred speech and staggering gait, and of the cortex, leading to impairments in performance of a variety of cognitive tasks, for example, reaction time tests, memory deficits, and, finally, intellectual functioning. At moderate doses, the behavioral responses to alcohol vary widely among individuals. This is because the social setting and an individual's mental state are important determinants of the effect of the drug. In one situation, alcohol intake may be associated with euphoria and relaxation, while in another, it may lead to withdrawal, hostility, and aggression. Setting and state of mind become progressively less important with higher doses, since the depressive actions of alcohol predominate leading to difficulties in talking, walking, and thinking.

Alcohol decreases alertness, impairs reaction time, leading to disturbances in motor skills, and negatively affects the performance of intellectual tasks (Zeigler et al. 2005). Alcohol amnesia and black outs, after which the drinker cannot remember the events which occurred while he or she was intoxicated, are not uncommon among heavy drinkers, especially those who engage in binge drinking behavior. Consumption of very large quantities of alcohol leads to severe depression of brain activity, producing stupor, coma, and ultimately death resulting from inhibition of the respiration centers in the brain stem. Lethal levels of alcohol range from 0.4% to 0.6% by volume in the blood. However, most individuals pass out before drinking an amount of alcohol capable of causing death (Nelson et al. 2004).

11.2.2   Chronic Alcohol Use

Both peripheral and CNS damage occur in the case of chronic alcohol abuse (Schweinsburg et al. 2003). Peripheral nerve damage is characterized by numbness in the feet, muscle discomfort, and fatigue, particularly in the lower part of the legs. With continued drinking, the alcoholic may experience weakness in the toes and ankles, loss of vibratory sensations, impairments in the control of fine movements, and ultimately decreased pain sensitivity. Alcohol elicits direct neurotoxicity and damage to the CNS and is one of the most devastating consequences of chronic and large alcohol intake (Suter 2012). This damage is frequently manifested by the appearance of deficits in cognitive capacity. 50 to 70% of detoxified alcoholics display lower performance on tests of learning, memory, and perceptual motor skills than their nonalcoholic counterparts (Zeigler et al. 2005). As the deleterious effects of chronic alcoholism are well documented, there is recent controversy that suggests moderate alcohol consumption to enhance cognitive capacity in old age. The first indications of alcohol-induced neurotoxicity, or what is termed Wernicke's encephalopathy, are disturbances in balance and gait, weakness of the eye muscles, difficulties with short-term memory, and mental confusion. Continued consumption of alcohol can lead to Wernicke-Korsakoff syndrome, a persistent neurological disorder typified by an inability to form new memories (anterograde amnesia), a disordered sense of time, hallucinations, confabulation, and dementia.

11.2.3   Alcohol-Influenced Depression

The link between alcohol consumption and the prevalence of depression is an area of interest, as has been demonstrated by recent research. A meta-analysis of 13 epidemiological investigations published since 2001 (Boden and Fergusson 2011) demonstrated an association between alcohol use disorder (AUD) and major depression (MD), with a further attempt to determine whether AUD causes MD or whether MD causes AUD or whether AUD and MD simply co-occur or whether a reciprocal causal relationship exists between the disorders whereby each simultaneously increases the risk of the other, was carried out. They concluded that a causal link appeared to exist between AUD and MD such that increasing use of alcohol increased the risk of depression. To explain the causal link, the authors suggested three possible mechanisms: (1)that alcohol exposure may cause metabolic changes that also act to increase the risk for depression, for example, reducing the enzymatic activity related to folate metabolism (McEachin et al. 2008); (2) that AUD and MD may be linked by genetic factors which relate to neurotransmitter functioning, so that the risk of depression is increased when alcohol is available (Saunders et al. 2009); and (3)that the abuse of alcohol is associated with a variety of issues including disruptions in family life, employment, and legal difficulties which lead to depression (Boden and Fergusson 2011).

11.3   Interrelationship Between Nutrition and Alcohol-Induced Neurotoxicity

The ways in which alcohol produces its neurotoxic effects remain unclear. However, there is growing evidence that nutritional deficiencies play a significant role in alcohol-induced brain damage. Particular attention has been paid to thiamin because of the critical role this vitamin plays in brain metabolism. Thiamin serves as a cofactor for a number of enzymes important in maintaining glucose metabolism in nerve cells. In thiamin deficiency, there is a reduction in the production of these enzymes which consequently results in impairment of the energy-producing abilities of the cells in the brain. In about 80% of alcoholics, thiamin deficiency occurs. Additionally, alcohol brings about the following to internal organs:

Damages the liming of the intestine, which decreases the intestinal absorption of thiamin
  Reduces metabolism of thiamin
  Impairs storage of thiamin within the liver (Subramanya et al. 2010).

Investigations which demonstrate that thiamin deficiency, resulting from gastrointestinal disease, gastric resection, anorexia nervosa, and other disease states, leads to neurological and behavioral outcomes essentially the same way as those that occur in Wernicke-Korsakoff syndrome shows how important thiamin deficiency is as an etiological factor in alcohol-induced neurotoxicity. In addition, both research and clinical studies have demonstrated that administration of thiamin can reverse the early symptoms of Wernicke-Korsakoff syndrome. Though a large percentage of alcoholics suffer thiamin deficiency, only a small (10%) develop the Wernicke-Korsakoff syndrome. One possible explanation for this fact is that genetic factors play a part in mediating the effects of thiamin deficiency on brain function and behavior. Evidence for this comes from animal studies showing differences among inbred strains in susceptibility to the neurotoxic effects of thiamin deficiency. It is then safe to conclude that there are potential benefits in treating alcoholics with thiamin to reverse the Wernicke-Korsakoff syndrome, though the exact amount needed to achieve this is controversial. In addition, as most alcoholics suffer from a variety of nutrient deficiencies, it is important to note that thiamin alone may be insufficient to ameliorate the mental problems associated with alcohol abuse. Finally, it must be recognized that if brain damage is extensive, nutritional therapy will be unable to reverse the adverse effects of prolonged alcohol abuse.

11.4 The Effect of Alcohol on Fetal Development

While the brain of an alcoholic is severely affected, it is important to note that it is the brain of a developing fetus that is equally, if not more, affected by alcohol intake by the mother. Alcohol very severely affects the developing fetus in a mother's womb. It has long been suspected that imbibing alcohol during pregnancy can affect the developing fetus. Legend has it, that, on the wedding night alcohol was forbidden to the young wife for fear of birth of a damaged child. This has been part of the western culture. Aristotle frequently warned the women drunkards of the birth of a defective child. In England, the "gin epidemic" which occurred between 1720 and 1750 strengthened the suspicion that alcohol could harm fetal development. This epidemic was the direct consequence of the availability of cheap food imports in the early eighteenth-century England and the consequent decrease in the demand for domestic grain. The British Parliament promoted distilling native grain to make alcohol and removed taxes on gin, the resulting alcoholic beverage. As alcohol abuse became rampant, the number of infants born with developmental and behavioral defects soared. A report by the Royal College of Physicians portrayed the offspring of alcoholic mothers as "starved, shriveled and with imperfect look." In 1751, to protect the health of the mothers and infants, the British Parliament reinstated taxes on gin and other distilled beverages (Abel 1984).

The importance of the interrelationship between maternal alcohol consumption and birth defects in the children born to such mothers was recognized as an important area of research only from the beginning of twentieth century. Children born to alcoholic mothers had higher mortality and morbidity rates compared to those born to the nonalcoholic mothers (Sullivan 1989). Unfortunately, the scientific fraternity did not recognize this as an important area of research, and but little was published on the adverse effects of alcoholism on childbirth during the next 70 years (Abel 1984). It is from 1968 that the modern era of research began on the effects of maternal alcohol consumption and fetal development. The physical and behavioral aspects of children born to alcoholic mothers were first published in France by Lemoine et al. (1968). Unfortunately, since the paper was published in French, it received little attention in the USA. Five years later, publication of two research papers provided a more detailed picture of growth failure, dysmorphic facial features, and neurological abnormalities in the offspring of alcoholic mothers (Jones et al. 1973). It was in 1973 that Jones and Smith (1973) coined the term fetal alcohol syndrome (FAS) to describe the constellation of abnormalities resulting from maternal alcohol abuse. The classic characteristics of FAS include pre- and postnatal growth retardation, a pattern of recognizable facial abnormalities, and complex alterations in brain structure and function, which are associated with impairments in both cognitive and social behaviors. From the start, it was recognized that the classic triad of features of FAS, namely, growth retardation, facial changes, and severe cognitive dysfunctions, are actually relatively uncommon in the offspring of women who drink heavily during pregnancy. More frequently, children of these women who drink heavily manifest more subtle signs of brain injury and cognitive deficits, with or without, the physical features of FAS. Thus, in 1996, the Institute of Medicine added definitions for alcohol-related birth defects (ARBD) and alcohol-related neurodevelopment disorder (ARND) to the array of alcohol-related birth defects (Institute of Medicine 1996).

11.5 Fetal Alcohol Syndrome (FAS) and Its Behavioral Consequences

Of the several consequences of excess drinking by expectant mothers, the most damaging is the impairments on the behavior of the infant that is born to these mothers. Ironically, mental retardation associated with alcohol intake is entirely preventable. Children with FAS vary widely in their cognitive and intellectual capabilities, but most suffer from mild to moderate mental retardation, exhibiting IQ scores which range from 20 to 120, with a mean of 65 (Walker and Johnson 2006). The severity of the intellectual deficits is directly related to the degree of physical anomalies in children with FAS. Children exposed to alcohol during development not only suffer from global mental retardation but also display a variety of more specific cognitive defects. Children with FAS and ARND exhibit impairments in their organizational skills; ability to learn, from previous experiences, and to foresee consequences; and speech, language, and communication capabilities (Koren et al. 2003). Additionally, prenatal exposure to alcohol is linked to a multiplicity of behavioral abnormalities. Infants diagnosed with FAS are irritable and jittery and often display low levels of arousal, abnormal reflexes, and a decrease in habituation to repetitive stimuli (Sokol et al. 2003). Many infants with FAS have a weak sucking reflex, leading to poor feeding and consequent developmental growth failure during the postnatal period. Sleep disturbances are also common in infants exposed to high doses of alcohol consumption by the mother during fetal development. In older children, hyperactivity and lowered attention are hallmarks of prenatal exposure to alcohol. Long-term follow-up investigations of children with alcohol-related disorders reveal that problems of hyperactivity persist throughout childhood and may continue well into adulthood (Wass et al. 2002). In addition, children with FAS and related disorders routinely display impairments in both gross and fine motor behavior (Korkman et al. 2003). Recent investigations indicate that difficulties in social situations frequently plague individuals affected by FAS. In infancy, the mother-child relationship can be affected, and, when children with FAS enter school, they display a higher level of aggression and have the social skills of children younger than their chronological age (Coggins et al. 2003). Unfortunately, the FAS does not improve with time and remains a permanent fixture in these individuals. These deficiencies in social behavior can impair work performance and increase the possibility of the individuals with FAS who will need to depend on their families and other social agencies throughout their lives.

11.6   Drinking Alcohol: The Eternal Question "How Much Is Too Much" in FAS-Affected Individuals?

This is one of the most frequently asked questions, and, unfortunately, the answer is simply "We do not know." Research suggests that even low levels of prenatal exposure to alcohol can have adverse impact on the fetus. Even a single drink per day leads to the birth of FAS-affected infants. And the relationship with the quantum of alcohol consumed is directly related to the severity of the symptoms. The more a mother drinks during pregnancy, the greater the possibility that she will bear a child with FAS or other morphological or behavioral symptoms associated with alcohol abuse. Thus, the question how much alcohol is safe to drink during pregnancy is a question which remains unanswered to this day. Although an occasional drink might not exhibit severe FAS, even low levels of maternal alcohol consumption during pregnancy can lead to behavioral complexities in the infants born. Thus, it is better to err on the side of caution. One can seek advice from agencies, such as the Department of Health in the UK or Centers for Disease Control and Prevention in the USA. The safest course, of course, is total abstinence during pregnancy or when one plans to have a child (Suter 2012).

References

Abel, E. L. (1984). *Fetal alcohol syndrome and fetal alcohol effects*. New York: Plenum Press.

Boden, J. M., & Fergusson, D. M. (2011). Alcohol and depression. *Addiction, 106*, 906–914.

Cantrell, P. A. (2000). Beer and ale. In K. F. Kiple & K. C. Ornelas (Eds.), *The Cambridge world history of food* (pp. 619–625). Cambridge: Cambridge University Press.

Caton, S. J., Ball, M., Aherm, A., & Hetherington, M. M. (2004). Dose-dependent effects of alcohol and appetite and food intake. *Physiology and Behavior, 81*, 51–58.

Coggins, T. E., Oiswant, L. B., Olson, H. C., & Timler, G. R. (2003). On becoming social competent communicators: The challenge for children with fetal alcohol exposure. *International Review of Research in Mental Retardation, 27*, 121–150.

Diagnostic and Statistical Manual of Mental Disorders. 5th edn (DSM-5). (2013). American Psychiatric Association, Washington, DC.

Gronbaek, M. (2004). Epidemiologic evidence for the cardioprotective effects associated with consumption of alcoholic beverages. *Pathophysiology, 10*, 83–92.

Institute of Medicine. (1996). Division of biobehavioral sciences and mental disorders. Committee to study fetal alcohol syndrome. In K. Stratton, C. Howe, & F. Battaglia (Eds.), *Fetal alcohol syndrome: Diagnosis epidemiology, prevention, and treatment*. Washington, DC: National Academy Press.

Jones, K. L., & Smith, D. W. (1973). Recognition of the fetal alcohol syndrome in early infancy. *The Lancet, 302*, 999–1001.

Jones, K. L., Smith, D. W., Ulleland, C. N., & Streissguth, A. P. (1973). Pattern of malinformation in offspring of chronic alcoholic mother. *The Lancet, 301*, 1267–1271.

Koren, G., Nulman, I., Chudley, A. E., & Loocke, C. (2003). Fetal alcohol spectrum disorder. *Canadian Medical Association Journal, 169*, 1181–1185.

Korkman, M., Kettunen, S., & Autti-Ramo, I. (2003). Neurocognitive impairment in early adolescence following prenatal alcohol exposure of varying duration. *Child Neuropsychology, 9*, 117–128.

Lemoine, P., Haronsseau, H., Borteyu, J. P., & Menuet, J. C. (1968). Les enfants de patents alcoolique observes a propose de 127 cas. *Qust Medcical, 25*, 476–482.

McEachin, R. C., Keller, B. J., Saunders, E. F., & McInnis, M. G. (2008). Modeling gene-by-environment interaction in comorbid depression with alcohol use disorders via an integrated bio-informatics approach. *BioData Mining, 1*, 2.

Mokdad, A. H., Marks, J. S., Stroup, D. F., & Gerberding, J. L. (2004). Actual causes of death in the United States. *Journal of the American Medical Association, 291*, 1238–1245.

Nair, K. P. P. (1996). The buffering power of plant nutrients and effects on availability. *Advances in Agronomy, 57*, 237–287.

National Institute on Alcohol Abuse and Alcoholism. (2014). *Alcohol facts and statistics*. Available at: http://www.niaaa.nih.ov/alcohol-health/overview-alcohol-consumption/alcohol-facts-and-statistics. Accessed 25 Mar 2015.

Nelson, E. C., Heath, A. C., Buchholz, K. K., Madden, P. A. F., & Fu, Q. (2004). Genetic epidemiology of alcohol-induced blackouts. *Archives of general Psychiatry, 61*, 257–263.

Newman, J. I. (2000). Wine. In K. F. Kiple & K. C. Omelas (Eds.), *The Cambridge world history of food* (pp. 730–737). Cambridge: Cambridge University Press.

O'Shea, R. S., Dasarathy, S., & McCullough, A. J. (2010). Alcoholic disease. *Hepatology, 51*, 307–328.

Saunders, E. E., Zhang, P., Copeland, J. N., McInnis, M. G., & Zollner, S. (2009). Suggestive linkages at 9p22 in bipolar disorder weighted by alcohol abuses. *American Journal of Molecular Genetics Part B: Neuropsychiatric Genetics, 150B*, 1133–1138.

Schweinsburg, B., Alhassoon, O., Taylor, M., Gonzalez, R., & Videen, J. S. (2003). Effects of alcoholism and gender on brain metabolism. *American Journal of Psychiatry, 160*, 1180–1183.

Sokol, R. J., Delaney-Black, V., & Nordstrom, B. (2003). Fetal alcohol spectrum disorder. *Journal of the American Medical Association, 290*, 2996–2999.

Subramanya, S. B., Subramanian, V. S., & Said, H. M. (2010). Chronic alcohol consumption and intestinal thiamin absorption: Effects on physiological and molecular parameters of the uptake process. *American Journal of Physiology: Gastrointestinal and Liver Physiology, 299*, G23–G31.

Sullivan, W. C. (1989). A note on the influence of maternal inebriety on the offspring. *Journal of Mental Science, 45*, 489–503.

Suter, P. M. (2012). Alcohol: Its role in nutrition and health. In J. W. Erdman, I. A. Macdonald, & S. H. Zeisel (Eds.), *Present knowledge in nutrition* (10th ed., pp. 912–938). Ames: Wiley-Blackwell.

Thomasson, H. R. (1995). Gender differences in alcohol metabolism: Physiological responses to ethanol. In M. Galanter (Ed.), *Recent developments in alcoholism* (Vol. 12, pp. 163–179). New York: Plenum Press.

Walker, W. O., & Johnson, C. O. (2006). Mental retardation: Overview and diagnosis. *Pediatrics in Review, 27*, 204–212.

Wannamethee, S. G., & Shaper, A. G. (2003). Alcohol, body weight, and weight gain in middle-aged men. *American Journal of Clinical Nutrition, 77*, 1312–1317.

Wass, T. S., Simmons, R. W., Thomas, J. D., & Riley, E. P. (2002). Timing accuracy and variability in children with prenatal exposure to alcohol. *Alcoholism: Clinical and Experimental Research, 26*, 1887–1896.

Zeigler, D. W., Wang, C. C., Yoast, R. A., Dickinson, B. D., & McCaffree, A. (2005). The neurocognitive effects of alcohol on adolescents and college students. *Preventive Medicine, 40*, 23–32.

Chapter 12
Anorexia Nervosa and Bulimia Nervosa: Two Most Important Eating Disorders of the Millennium

Abstract The chapter, principally, discusses, at great length, the two most important eating disorders of the millennium anorexia nervosa and bulimia nervosa, the former emanating from a fear of eating unfamiliar foods, leading to self-imposed starvation, and the latter to "binge eating." A historical perspective of these diseases is also discussed. The chapter would also discuss how to tackle these eating disorders, with the least negative mental impact on the patients.

Keywords Anorexia nervosa · Bulimia nervosa · Eating disorders · Psychological treatment

It is lately that a growing interest is seen in two eating disorders known as anorexia nervosa (see image below) and bulimia nervosa. Eating disorders vary widely, ranging from the fear of eating unfamiliar foods (food neophobia), to the eating of non-food substances (pica), to compulsive overeating (binge eating). Self-imposed starvation can lead to death, as has been highlighted by the deaths of Karen Carpenter, an American pop singer in 1983, and Christy Henrich, an Olympic gymnast in 1994. Over the past decade, the rash of fashion models who have died due to complications of anorexia, namely, Ana Carolina Reston, Hila Elmalich, and Isabelle Caro, in the USA, to name but a few, has brought increased scrutiny to the unbearable pressures which are apparently placed on young women to appear beautiful. In a world where one in every five persons experiences chronic hunger, and where millions starve or sleep each day with half-filled bellies, like in India – despite a so-called green revolution which was supposed to have made the country "self-sufficient" in food – it is ironical that certain individuals would willingly attempt to starve, when sufficient food is available for them to consume, in order just to look good. It is in western countries that an irrational fear of fatness by young women, in particular, has made the disorders of anorexia and bulimia nervosa a topic of great concern to the medical, psychological, and nutritional communities.

© Springer Nature Switzerland AG 2020
K. P. Nair, *Food and Human Responses*,
https://doi.org/10.1007/978-3-030-35437-4_12

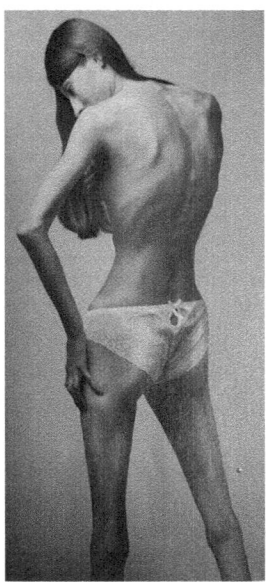

A young woman suffering from anorexia nervosa. (*Source:* Shutterstock)

12.1   Anorexia Nervosa

The eating disorder starts quite innocently. Look at the following example. A teen age girl may start cutting on sweets, snacks, etc. and is praised by her mother for her will power. Even after reaching the desired weight, she will continue with the "dieting," limiting her diet just to fruits and vegetables and reach a situation where her daily intake is no more than 600 calories. When a friend confronts her about her excessive weight loss, she complains that either she is still too heavy or some parts of her body are excessively fat. Anorexia nervosa is truly an unusual eating disorder. Although individuals suffering from this eating disorder may show many symptoms which are seen in other forms of starvation, they have one unique characteristic – *their hunger is deliberate and self-imposed.* In a manner, the disorder represents a paradox, as it is most often seen in young females who are considered well-off, yet who are starving amidst plenty. The term anorexia nervosa even appears inappropriate to explain the condition, because it derives from the root "an-orexis-nervosa" (without appetite of nervous origin) and suggests the victim has no appetite. But, individuals with this disorder do not suffer from a loss of appetite; rather, they suffer from an intense fear of gaining weight. The term *hypophagia* would be more appropriate (Booth 1994), to describe the disorder, if a Greek-based name is desired, as it means "deficient eating." Nevertheless, individuals with anorexia nervosa are frequently preoccupied with eating and other thoughts of food. This is the paradox of this eating disorder, because, unless one is constantly preoccupied with thoughts of food, and what might follow if one were to eat too much, why would one starve at all?

12.2   Anorexia Nervosa: A Historical Perspective

There is something strange with this eating disorder, because, religion comes in between. It is interesting to note that historians who study religion have described the pious behavior of several saints as resembling what we now view as anorexic. Saint Catherine of Siena, for instance, who lived in the fourteenth century, led a life of asceticism in order to garner spiritual fulfillment, barely subsisting on a diet which lacked in both quality and quantity. To expiate sin, she abstained from food and drink, except taking the communion supper, and, eventually, died of starvation. Nearer home, in India, there is an age-old practice of abstaining from food, called *Upavasa* in Sanskrit (the ancient language of India); it means abstaining from food. More pictorially, late Mahatma Gandhi took to *Satyagraha* (again a Sanskrit word meaning the pursuit of truth) when he had to fight the British, by abstaining from food, except drinking water. It is also interesting that during a specific religious period in India (like the 9 days, called *Navaratri*, in India), when the goddess of knowledge, *Saraswati*, and *Durga*, the one who demolished a demon, which is annually celebrated in the month of October, pious people abstain from food for 9 days. Famously, the Indian Prime Minister, Shri Narendra Modi, a staunch Hindu, abstained from food for full 9 days during his visit to the USA, in 2014, immediately after his election, on the invitation of former US President Barack Obama, sipping only lemon juice and warm water. The term "holy anorexia" to describe the health disorder of Saint Catherine of Siena was coined by Robert Bell, though he points out that cultures in which such women lived valued spiritual health and fasting rather than starvation to "look thin and beautiful" as modern-day women in the west do. In the early 1870s, European physicians offered the first "modern" description of anorexia. In an address he delivered at Oxford, Sir William Gull (Gull 1874) referred to " a particular form of disease occurring mostly in young women characterized by extreme emaciation …The subjects of this affection are mostly of the female sex and chiefly between the ages of 16 and 23. I have occasionally seen it in males at the same age…. The condition was one of simple starvation…. The want of appetite is due to a morbid mental state…We might call the state hysterical……I prefer however the more general term nervosa."

Gull was, in fact, the first to use the term "anorexia nervosa," and his description from well over a century ago still paints a fairly accurate portrait of what the disorder looks like. Thanks to his work, and also the reports of anorexia provided by Dr Charles Lasegue of France, substantial increase occurred in the medical community's awareness of the disease. With the advent of Freudian thinking in the first half of the twentieth century, a psychoanalytic perspective dominated clinicians' ideas about the disorder. Specifically, the fear of "incestuous impregnation" was at the core of the early drive theories of anorexia and served to explain the patient's disease in the oral intake of food. Although this idea may seem outrageous today, the psychoanalytic approach represented the first serious attempt to understand the etiology of the disorder.

Hilde Bruch, a psychoanalyst, began to study anorectic women in the 1950s, when the current wave of interest in anorexia nervosa actually began. Bruch suggested that the rigid behavior that these anorectic women associated with eating represented a struggle for control. As girls, they had been good children, who were quiet, obedient, clean, helpful, dependable, and eager to please and excelled in studies. At adolescence, when they were supposed to be self-reliant, they lacked an age-appropriate sense of autonomy and became overcomplaint and oversubmissive. Bruch viewed the rigid control of eating and the excessive concern with the body size as symptoms of their desperate fight against feeling enslaved, exploited, and incompetent. She proposed that their strict limitation of food intake might be one of the only ways the anorectic could feel in control of her life (Bruch 1973). By the 1970s, a dramatic rise in both medical and public interest in anorexia was evident. Researchers began exploring not just the etiological factors involved, but, physiological as well as psychological. With the high-profile death of Karen Carpenter, earlier referred to in this review, public interest snowballed. As celebrities like Victoria Beckham, aka Posh Spice, came out with their open confessions, in the 1990s, articles in the press, talk shows, TV shows, etc. boomed on the eating disorder.

12.3   What Are the Diagnostic Criteria for Anorexia Nervosa?

It was in 1980 that the American Psychiatric Association which recognized "anorexia nervosa" as a classifiable mental disorder with the 1980 edition of the Diagnostic and Statistical Manual of Mental Disorders, which has been updated as per the details shown in the following table (Table 12.1), that certain criteria were spelled out to define this eating disorder.

The most recent DSM-5 diagnostic criteria for anorexia nervosa distinguish among level of severity, based on the individual's body mass index (BMI, Table 12.2, below). BMI is determined using the mathematical formula of mass in kilograms divided by height in meters squared of the individual concerned.

Prior editions of the DSM also required that a female must stop menstruating in order to be classified as anorexic, or must have her menarche delayed, if she is prepubertal. This physiological disturbance has been attributed to primary nervous system dysfunction, but may be better explained by caloric intake restriction and the lowered fat content of the body (Pomeroy 1996). Presumably, this serves to protect the female from becoming pregnant, which would place even greater demands on her body. However, some debate surrounded the utility of this requirement (Watson and Andersen 2003), and it was deleted from the diagnostic criteria of DSM-5. Retained in the most recent DSM guidelines is the distinction between the restricting type of anorexic and binge eating/purging type. The restricting type relies on her ability to sustain self-control in not eating. Additionally, this type of anorexic will engage in frequent, regular, and vigorous exercise as an added means of displacing

Table 12.1 Criteria for anorexia nervosa

Criterion 1: Restriction of energy intake relative to the body's requirements leading to a body weight less than minimally normal or expected in the context of age, sex, developmental trajectory, and physical health
Criterion 2: Intense fear of gaining body weight or becoming fat, or persistent behavior which interferes with weight gain, even though the individual is underweight
Criterion 3: Disturbances in the manner in which one's body weight or shape is experienced, undue influence of body weight or body shape on self-evaluation, or persistent inability to recognize the seriousness of the current low body weight
In the *restricting type*, the individual has lost weight through persistent dieting, fasting, or exercising excessively. In the *binge eating/purging type,* the individual has engaged in recurrent episodes of binge eating or purging behavior

Source: DSM 5 – Diagnostic and Statistical Manual of Mental Disorders (2013)

Table 12.2 BMI percentile cutoffs for severity of anorexia nervosa

Details	BMI (kg/m^2)
Mild	17–18.49
Moderate	16–16.99
Severe	15–15.99
Extreme	<15

hunger as well as to work off calories from the minimal amounts of food that she has eaten. As bingeing and purging are a primary characteristic of the bulimic sufferer, these behaviors will be addressed in the section which discusses bulimia nervosa.

Anorexia nervosa prevails most among adolescent girls clocking 85–95%. It has been, for long, believed that the disorder prevails most among the upper class, but that is not true anymore. It diffuses through the population cutting across social classes. Further, it was believed that the disorder prevailed only among the whites and not in African-American women. This is no more true, but, on a relative basis, it prevails more among the black community than in the Caucasian community. How culture may indirectly influence eating attitudes was shown in a study conducted in 1983. The researchers compared three groups of women: British Caucasian, Kenyan, and Kenyans living in Britain. The Kenyans preferred a larger body shape than the British Caucasians, while, very surprisingly, the Kenyans who lived in Britain preferred the thinnest shape. With respect to body image, the immigrants appeared to overassimilate. An interesting study is from South Africa. Prior to the 1990s, white South African schoolgirls were indistinguishable from their European and American counterparts in terms of their overvaluation of thinness and vulnerability to eating disorders seen among black South Africans. Since the change from apartheid to democratic rule, however, studies of black South African collegiate women reveal scores at least as high on disordered eating measures as their Caucasian classmates (Wassenaar et al. 2000). Assimilation may be a partial explanation, but it remains to be seen just what contribution race may make to the likelihood of an eating disorder.

12.4   Anorexia Nervosa: Its Physiological Consequences

Numerous medical complications follow with the onset of this disorder. They are summarized in the following table (Table 12.3).

The medical complications following anorexia nervosa are numerous as detailed above (Table 12.3). Many of the medical complications are similar to those which accompany other forms of starvation, but others do appear to be specific manifestations of anorexia nervosa (Table 12.3). As is but to be expected, the physiological complications that appear in anorexia nervosa vary directly as a function of the degree of reduced food intake and body weight.

Of all symptoms, the cardiovascular complications are the severest in anorexia nervosa. In fact, the standardized mortality rate for anorexia nervosa is estimated to be around 9.6% – a staggering figure for a psychiatric disorder, with one long-term study putting it at 18% (Hoek 2006).

The metabolic changes associated with the disorder are similar to those seen with starvation. After a number of days of semi-starvation, glycogen reserves are first depleted, and then protein oxidation increases until ketogenesis kicks in, at which point fat oxidation starts. This can result in elevated free fatty acids and higher plasma cholesterol. If the individual continues to eat fewer calories than are required for the proper body needs, tissue may eventually be consumed in order to meet the energy deficit. Other metabolic problems associated with the disorder are abnormal vasopressin release, which results in increased urine output and concomitant reduction in intracellular water (Hill and Pomeroy 2001). Additionally, the disorder can also lead to electrolyte imbalances, anemia, and other mineral deficiencies, kidney stones if chronically dehydrated, and edema during the re-feeding process (Gary et al. 2012). All of these symptoms taken together illustrate the severe taxing of the body that self-starvation via anorexia nervosa inflicts on its victim.

Table 12.3 Consequences of anorexia nervosa

Details	Symptoms
Cardiovascular	Rapid and slowed heart rate, arrhythmia, hypotension, dizziness, fainting
Endocrine	Infrequent menstruation, amenorrhea, anovulation, cold sensitivity
Hematologic	Anemia, low white blood cell count, excess bleeding or bruising, fatigue and weakness, excess glucocorticoids
Integumentary	Dry, flaky, scaly, orange colored skin, decreased body fat, lanugo, thinning hair, brittle nails
Gastrointestinal	Abdominal discomfort, constipation, bloating, delayed gastric emptying, decreased gastric motility, pancreatitis
Skeletal	Lowered bone mineral density, osteoporosis, bone fractures, stunted growth
Fluids and electrolytes	Electrolyte imbalance, dehydration, peripheral edema, metabolic acidosis, renal failure
Central nervous system	Decreased concentration and attention, poor problem-solving and memory depressed mood, peripheral neuropathy, seizures

Source: Gary et al. (2012)

12.5   Psychological Fallout of Anorexia Nervosa

The psychological effects of anorexia nervosa are similar to those of starvation. Low tolerance for stress, irritability, and sexual disinterest but also a preoccupation with food and bizarre eating habits characterize both starvation and anorexia nervosa. Labile emotions, also an avoidance of emotional intimacy and social interaction, yet maintaining a dependence of others characterize the disorder (Woolsey 2002). A preoccupation with food manifests in preparing meals for others, collecting recipes, attending cooking classes, and also taking courses in nutrition and dietetics (Worobey and Schoenfeld 1999) which also are the other characteristics of the disorder. Bizarre eating habits may also include refusal to eat in the presence of others or taking an extremely long time to eat a small portion of food that is served. Although many of the psychological consequences observed in the individuals suffering anorexia nervosa and starvation are similar, there are several differences that make the disorder unique. For instance, anorexia tends to display a severe distortion in their body image (Grant et al. 2002). They will typically misperceive their body size as larger than they actually are, regardless of what the weighing machine indicates or the assurances of their trusted friends. Secondly, anorexics frequently misinterpret both internal and external stimuli. For example, internal cues for hunger are redefined as positively reinforcing, with less and less food necessary to feel full, perhaps an outcome of the delay in gastric emptying (Chial et al. 2002). Third, unlike starving individuals who become easily fatigued, and avoid physical activity, anorexics are often hyperactive and deny feeling fatigue (Hebebrand et al. 2003). Indeed, exercise may be further used to achieve weight loss. A fourth distinction relates to issues of control (Horowitz 2004). Anorexics feel helpless in most matters, but very strangely, they can control the eating urge; hence, their self-discipline with regard to eating urge is extraordinary. Finally, it is safe to assume that the fear of fatness which serves to motivate anorexic behavior has no equivalent in the situation of starvation.

12.6   The Bulimia Nervosa

This disorder is more widespread than anorexia nervosa, though the manifesting features are not as distinguishable as that of anorexia nervosa, and so it is more difficult to establish this disorder compared to anorexia nervosa. Though the disorder primarily occurs in women, more men suffer from it than women. While many women with bulimia may also suffer from anorexia, the average bulimic tends to be heavier and older than the typical anorexic. The disorder is defined by a binge-purge cycle of food intake. The disorder is more aptly named than anorexia nervosa, as the word bulimia derives from the Greek word "bous" (meaning ox) and "l'mos" (meaning hunger"), implying an oxen hunger with a ravenous appetite, which the binging behavior supports. Excessive amounts of calories are, indeed, ingested, but the individual compensates for the binge by purging through self-induced vomiting or other techniques. The primary distinction between the diagnoses of bulimia and anorexia may in fact be in the eating patterns, as bulimia is characterized by such frequent binges on food, while anorexia includes persistent efforts to restrict it (DSM-5 1013). Some bulimic sufferers may try to diet for a period of days or weeks between binges, but fail in their attempt and binge even more, while others may binge unremittingly.

It is unusual for a bulimic individual to eat a normal meal. If the individual is participating in a social function, and food is served lavishly, one would find a bulimic individual hardly eating. By comparison, during a binge episode, the amount of food equivalent to from 1500 calories to as high as high as 60,000 (by one estimate) will be consumed by a bulimic individual (Goldsmith 2006). Foods consumed during a binge tend to be high in fat and carbohydrate contents, as food is eaten for its emotional satisfaction not for its nutritional value. Hence, foods forbidden during dieting, like ice cream, chocolate, donuts, and milkshakes, top the list. Most binge eating is done in the evening and in secrecy. A variety of conditions can precipitate a binge episode, such as depression, stress, frustration, boredom, or just the plain sight of food. Following a binge, the individual typically feels further anxiety, depression, and now guilt. Due to humiliation because of having lost control or recognizing that all those calories will take their toll, the bulimic then seeks to undo his/her overeating. Most will regurgitate the food they ate, and many bulimics will anticipate vomiting in advance of their overeating. For other bulimics, purging is accompanied by using laxatives or diuretics. Unlike the anorexic, the individual with bulimia nervosa is aware of the abnormal nature of his/her behavior and thus attempts to hide it from his/her family and friends.

12.7   Historical Perspective of Bulimia Nervosa

It was the strange behavior of St. Mary Magdalene de'Pazzi that led the historian to speculate that the nun was, in fact, a bulimic, supposedly managing to lie on just bread and water, but, in secret gorging on food, as reported by some nuns. The abnormal gorging of food was attributed by the saint to possession by the devil. As early as 1890, clinical reports on bulimic binge eating were reported, but it was not until the 1940s that detailed reports of compulsive overeating and purging behavior were seen to appear in medical literature. It was not until the 1970s that reports on bulimic behavior among normal weight individuals became more available, with the surprising phenomenon observed in co-eds being attributed to the pressures of female socialization. Shortly thereafter, Gerald Russell used the term "bulimia nervosa" to explain the disorder. In the 1990s, the publicity garnered from admissions of bingeing and purging, by celebrities such as Princess Diana, Jane Fonda, etc., gave courage to thousands of women to come forth and seek understanding and treatment, with the sudden ascendence of bulimic syndromes described as spectacular.

12.8   Diagnostic Criteria for Bulimia Nervosa

Diagnostic criteria for bulimia nervosa are described in the following table (Table 12.4).

The self-esteem among bulemics is greatly influenced by their weight and body shape, as in the case of anorexics. But, unlike anorexics, most bulemics maintain a weight that is within normal limits. Women with bulimia typically follow an unsuccessful, dieting pattern to maintain weight loss, which becomes interspersed with bingeing and purging behaviors which result in weight fluctuations of 10 lb or more over short periods of time. Since normal weight is maintained, the individual may view purging as a quick and easy solution to the problem of excess calories. When we examine the physiological effects of bingeing and purging, however, the dire consequences of this cycle will be evident.

12.9   The Physiological Consequences of Bulimia Nervosa

The bulimic patient will superficially look a picture of health unlike the very thin anorexic. But, similar to anorexia, the bulimic patient will be affected as described in the following table (Table 12.5).

The symptoms can vary from minor problems, such as abrasions on the knuckles, from continued use of the hand to induce vomiting to life-threatening difficulties such as electrolyte disturbances resulting from continual purging (Mitchell and

Table 12.4   Criteria for bulimia nervosa

Recurrent episodes of binge eating, characterized by the following:
Eating in a discrete period of time, an amount of food that is definitely larger than what most people would eat, during a similar period of time, and under similar circumstances
A sense of lack of control overeating during the episode
Recurrent inappropriate compensatory behavior in order to prevent weight gain, such as self-induced vomiting, misuse of laxatives, diuretics, or other medications, fasting, or excessive exercise
The binge eating and inappropriate compensatory behaviors both occur on an average, at least once a week, for 3 months
Self-evaluation is unduly influenced by body shape and weight

Source: DSM-5 – Diagnostic and Statistical Manual of Mental Disorders (2013)

Table 12.5   Symptoms of bulimia nervosa

Anovulation, calluses on the knuckles, constipation, dental caries, diarrhea, dry, flaky skin, indigestion, inflammation of the esophagus, heart failure, heart muscle disease, hypotension, itching, metabolic acidosis, metabolic alkalosis, mitral valve prolapse, muscle cramps, palpitation, pancreatitis, scaling of the lips, skin disorders, sore throat, and vomiting blood

Source: Robert-McComb and McCullough (2012)

Crow 2010). Many of the medical complications of bulimia nervosa are side effects of disordered eating practices. The rapid consumption of a large quantity of food can lead to acute gastric dilation with resulting discomfort. Inflammation of the pancreas, abdominal distention and pain, and increased heart rate may develop as a consequence of abrupt pancreatic stimulation during frequent binge eating episodes. Most bulic patients vomit regularly after a binge episode in order to remove from the stomach what they have just eaten. However, recurrent vomiting of the stomach's acidic contents will result in erosion of the esophagus, which, if torn, can be a life-threatening event in itself. Dental erosion, gum problems, and swelling of the salivary glands are also common consequences of continual vomiting and can serve as a diagnostic indicator of bulimic behavior (Robert-McComb and McCullough 2012). Repeated vomiting can also lead to loss of fluid, dehydration, and electrolyte imbalances. Bulimic patients also experience excessive thirst and decreased urinary output, resulting in edema from excess water being retained. The loss of Na, K, and chloride from the body can lead to a variety of cardiac symptoms ranging from irregular heartbeat to congestive heart failure and cardiac death (Mitchell and Crow 2010). Mortality rate for bulimic patients is about 3.9% (Crow et al. 2009). Despite their frequent vomiting, bulimic women do not appear to develop taste aversions to the food they eat, when they binge. Some believe it is because the vomiting is self-induced rather than imposed externally or caused by illness. Others propose that taste receptors on the palate may be damaged from acid in the vomit; hence, taste sensitivity is decreased. However, recent research has not borne this fact out.

12.10   Psychological Characteristics of Bulimics

There is similarity between the bulimic and anorexic patients inasmuch as the psychological characteristics of the disorder are concerned. Like anorexics, the bulimics seem to have a distorted view of their own weight and shape and desire to weigh much less than they really do and appear to be controlling their food intake. But, while anorexics are successful in exercising control over food intake, bulimics will repeatedly fail in this challenge and end up eating uncontrollably. Their effort at purging what they have just eaten is their shameful recourse for not having had the strength to resist the temptation to eat. Compared to anorexics, bulimics are more likely to display symptoms of depression, show greater lability of mood, and are higher in negative affectivity (Stice et al. 2006). In contrast to anorexics, who restrict both their caloric intake and their sexual activity, bulimics tend to engage in more sexual behavior (Kluck et al. 2012). And, unlike the anorexic patient who tries to resist treatment because of the fear of weight gain, the bulimic patient is usually more embarrassed from having been caught at bingeing and purging than is frightened at the prospect of receiving help. Patients of bulimia have been characterized as high achievers, yet tend to be dependent, passive, self-conscious, and unassertive and have difficulty communicating their feelings. Moreover, they are extremely vulnerable to rejection and failure and may have problems with impulse control, notably expressed through shoplifting and overconsumption of alcohol or drugs.

12.11   Treating Eating Disorders

As already discussed, both anorexia nervosa and bulimia nervosa can turn out to be life-threatening eating disorders. The early detection and appropriate treatment are imperative in view of the high mortality rates and traumatic psychological consequences associated with these diseases. Because of the complexities of these disorders, a multidimensional treatment approach is usually attempted. The initial step is a correct diagnosis. A careful medical history, including inquiries about dietary habits, weight fluctuations, and menstrual irregularities, should be recorded. This is followed by a complete and thorough physical examination of the patient, which includes the blood test, thyroid and renal function tests, and electrocardiogram. The physical examination will also allow the detection of delayed secondary sexual characteristics, lanugo, and carotenemia. A psychiatric assessment of the individual and the patient's family should also be undertaken, which examines attitudes toward food and body image disturbance (Andrews 2012).

12.12   Treating Anorexia Nervosa

The first concern here is nutritional rehabilitation, which should be planned by a physician in consultation with a registered dietitian. Without correcting the starvation state and starvation-induced emotional and cognitive states, the patient will not benefit from the accompanying psychological treatment. Caloric increases must be individualized to correspond to the patient's physical and mental condition and must be gradual because rapid calorific replacement during re-feeding therapy can be complicated by life-threatening pancreatitis or cardiomyopathy. Normal feeding is preferable to tube feeding or total parental nutrition, except for rare emergencies (Treasure 2004). Any therapeutic plan should be individualized, but the patient is typically given 500–700 kilocalories (2092–2929 kJ)/day to start with, with a goal of gaining 1–2 lb/week (under 1 kg) (Cockfield and Philpot 2009).

Although an eating disorder unit specialized center is the ideal location for inpatient treatment of anorexics, there are some conditions when hospitalization might be necessary. Severe, or rapid weight loss, that is, weighing less than 75% of the ideal, is a major signal for requiring hospitalization, and approximately one-half of anorexics require hospitalization (Hill and Pomeroy 2001). Additionally, life-threatening physical complications such as severe electrolyte imbalances, cardiac arrhythmias and dehydration, suicidal tendency, or if the patient's social situation interferes with outpatient treatment, all these are justifiable reasons for hospitalization

(Gore et al. 2001). It is also necessary to advise the patient that her weight will be carefully monitored so that she does not gain weight "too fast" (Hill and Pomeroy 2001). As the patient adjusts to the feeding regimen, caloric intake is slowly increased by about 200–300 kilo calories every other day (Cockfield and Philpot 2009). A critical feature of effective nutritional management is a therapeutic staff of nurses and a dietitian who can provide the patient with encouragement and support in her efforts to confront her ambivalence and compulsions. The nutritional program may require bed rest until a stable weight gain is evident, informational feedback about weight and calories, and gradual assumption of control eating by the patient. Once the patient begins to gain weight and appears to be motivated, psychological treatment can follow.

12.13   Tackling Bulimia Nervosa

For this, nutritional counseling may take the form of healthy meal planning in order to normalize eating patterns. Hospitalization is rarely necessary, although inpatient monitoring may be recommended if the patient's medical condition is unstable because of recurrent bingeing and purging or the abuse of laxatives or diuretics which create electrolyte imbalances. The treatment team must lay a solid foundation for nutrition education for the patient, help prevent the sequence of starvation, hunger, and overeating, and address her distorted body image and fears of weight gain.

12.14   Psychological Treatment for Anorexics and Bulimics

Though the external symptoms of these disorders are on the eating pattern, the origin of anorexia nervosa and bulimia nervosa is psychological in nature. Hence, the primary approach in treating these disorders has to be rooted in good psychology. What may be surprising, however, given the more recent identification of bulimia nervosa, as compared to anorexia nervosa, is that a greater number of well-designed investigations of treatment for bulimia exist than for anorexia. This may be, in part, due to the danger inherent in placing anorexic patients who are at physical risk in a "no-treatment" control group for the purposes of an investigation (Gleaves et al. 2000). Even designs where patients receiving treatment are compared to those who are waiting for treatment may be unethical. For example, in a randomized controlled trial of treatment for anorexia and bulimia, three patients in the control group became acutely ill while awaiting their deferred treatment. No one type of psychotherapy is ideal for treating anorexics and bulimics, as even the best therapeutic approaches fall short of helping all patients. Rather, psychotherapeutic strategies must be tailored to the cognitive styles, psychological profiles, and developmental levels of the participants. As each patient is truly unique, it is perhaps appropriate that a wide variety of therapies for the treatment of eating disorders are formulated

Table 12.6 Various psychotherapeutic approaches for the treatment of eating disorders

Biopsychosocial model, cognitive behavior therapy, psychodynamic therapy, feminist psychodynamic psychotherapy, interpersonal therapy, family therapy, marital therapy, psychoanalysis, focal psychoanalytic psychotherapy, dialectic therapy, supportive psychotherapy, psychoeducational therapy, pure/guided self-help and expressive therapy

Source: Santuccis (2014)

throughout the years. The following table (Table 12.6) provides a summary of these approaches.

As shown in the data in Table 12.6, there are well over a dozen distinct approaches which are consistently espoused, though some appear to be more effective than the others (Santuccis 2014). The treatments for anorexia and bulimia which have received the most empirical research, and, arguably the most effective, are family therapy, interpersonal therapy, and cognitive behavioral therapy, which will be discussed below.

Originally developed for the treatment of anorexia, family therapy (FT) has also been extended to treating bulimics. Family therapists argue that the eating disorder may have a distinctive role in precipitating or maintaining certain dysfunctional alliances, conflicts, or interaction patterns within the girl's family (Simic and Eisler 2012). Additionally, the eating disorder may serve as a maladaptive solution to the girl's struggle to achieve autonomy while providing a means by which the parents and child can avoid major conflict. By observing the family, issues of communication, loyalties, separation and individuation, enmeshment, over protectiveness, and conflict resolution, all of which are brought under control, and adequately dealt with. The emphasis is not on placing blame, but on identifying problematic functions within the family which may be hindering the patient's recovery (Gleaves et al. 2000).

For girls of age 18, or older, who have had anorexia symptoms for a relatively short period, family therapy (FT) has been found quite effective (Bulik et al. 2007). Therapy which addresses family communication, conflict, and cohesion appears to be beneficial to help anorexics to gain weight. These results suggest FT to be effective with bulimics as well, although its appropriateness as the sole treatment for bulimics merit further investigation (Gore et al. 2001). The focus of treatment in interpersonal therapy (IPT) is on categories of events which trigger problematic behaviors, such as emotion regulation, interpersonal role disputes, and interpersonal deficits (Good heart et al. 2012). IPT has shown some promise in treating bulimia in adults, being comparable to behavior therapy in reducing dietary restraint, and effective in decreasing patient's concerns about their body shape. Application of this procedure among children with bulimia and adolescents is also promising (Gleaves and Latner 2008). There is, yet, another third approach, cognitive behavior therapy (CBT), which combines behavioral techniques, such as self-monitoring of food intake and stimulus control, with cognitive strategies designed to combat dysfunctional thoughts about food and weight (Abbott and Goodheart 2012). Techniques for coping and preventing relapses are also typically included. It is the most widely investigated psychotherapeutic approach for the treatment of bulimics, but, given the greater difficulties in treating anorexics, there is not an established database for

Table 12.7 Phases of cognitive behavior therapy (CBT)

Phase 1
Build a therapeutic alliance
Establish the role of the patient as an active collaborator in treatment
Assess the core patterns and features of the patient's particular eating disorder
Orient the patient to the CBT model of eating disorders
Educate the patient about the importance of homework in CBT
Establish monitoring procedures for eating and weight changes
Eliminate dieting and normalizing eating throughout the day
Develop delay strategies and alternatives to binge eating and purging
Learn the skills of imagery, rehearsal, and relaxation training
Challenge assumptions about the overvaluation of shape and weight changes
Provide psychoeducational information to the patient
Phase 2
Eliminate forbidden foods
Identify dysfunctional thoughts
Cognitive restructuring
Problem-solving
Continue to challenge assumptions overvaluing weight and shape
Broaden the focus of treatment by identifying and challenging the eating disorder and self-schema and developing alternative bases for self-evaluation
Phase 3
Review high-risk situations and specific strategies to prevent relapse
Promote adaptive coping skills
Facilitate recovery from slips

Source: Pike et al. (2001)

measuring its efficacy with the latter. The data in Table 12.7 below show that it is highly structured, with 20 sessions spread over a few months for bulimics and 50 sessions for anorexics (Pike et al. 2001).

CBT has been found to decrease bulimic and depressive symptomatology, increase overall adjustment, and improve body image adjustment, with these improvements being maintained for at least 1 year after treatment (Gore et al. 2001). Although the mechanism by which CBT achieves its success is not entirely understood, it is the changing beliefs of the patients and behavioral pattern, which are thought to be crucial. One theory suggests that changing beliefs about the body will lead to changes in eating pattern, while another suggests that changing the eating and purging behavior is what leads to feelings of self-efficacy (Wilson et al. 2002).

12.15   Pharmacological Treatment

Although the foregoing discussion emphasized behavioral approaches to the treatment of eating disorders, a passing mention should be made of the relevance of pharmacological interventions which have been employed either in addition to or instead of psychotherapy. Early efforts to treat anorexics, for example, were based on the idea of treating the delusional beliefs of the anorexic patient concerning body weight and image. Hence, drugs that had been developed to treat psychosis, such as L-DOPA and lithium, were used with limited success (Kaplan and Howlett 2010). More recently, drug treatments have been derived from basic research on the mechanisms of food intake and the regulation of body weight. To date, various appetite stimulants, neuroleptics, opioid antagonists, and antidepressants have been tried, but none has shown to treat the core psychopathology of the disorder (Hoes and Curtis 2012). Since the symptoms of depression are also common with bulimics, researchers were quick to initiate studies on the effectiveness of antidepressants in ameliorating depression and anxiety in bulimics, as well as in altering their feeding behavior (Broft et al. 2010).

The principal limitation of pharmacological interventions in treating anorexics and bulimics is that their effectiveness may be short-lived. Patients have been known to relapse after medication is terminated, though a longer period of treatment may lead to longer success. However, some investigations have shown that even if kept on medication, one-third of bulimics may relapse (Romano et al. 2002).

References

Abbott, M., & Goodheart, K. L. (2012). Cognitive behavioral approaches for treating eating disorders. In K. L. Goodheart, J. R. Clopton, & J. J. Robert-McComb (Eds.), *Eating disorders in women and children* (pp. 371–383). Boca Raton: CRC Press.

Andrews, H. (2012). The assessment of mental state, psychiatric risk, and co-morbidity in eating disorders. In J. R. E. Fox & K. P. Goss (Eds.), *Eating and its disorders* (pp. 11–27). Malden: Wiley.

Booth, D. A. (1994). *Psychology of nutrition.* London: Taylor & Francis.

Broft, A., Berner, L. A., & Walsh, T. B. (2010). Pharmacotherapy for bulimia nervosa. In C. M. Grilo & J. E. Mitchell (Eds.), *The treatment of eating disorders: A clinical handbook* (pp. 388–401). New York: Guilford Press.

Bruch, H. (1973). *Eating disorders: Obesity, anorexia nervosa, and the person within.* New York: Basic Books.

Bulik, C. M., Berkman, N. D., Brownley, K. A., Sedway, J. A., & Lohr, K. N. (2007). Anorexia nervosa treatment: A systematic review of randomized controlled trials. *International Journal of Eating Disorders, 40,* 310–320.

Chial, H. J., McAlpine, D. E., & Camilleri, M. (2002). Anorexia nervosa: Manifestations and management for the gastroenterologist. *The American Journal of Gastroenterology, 97,* 255–269.

Cockfield, A., & Philpot, U. (2009). Feeding size 0: The challenges of anorexia nervosa. Managing anorexia from a dietitian's perspective. *Proceedings of the Nutrition Society, 68,* 281–288.

Crow, S. J., Peterson, C. B., Swanson, S. A., Raymond, N. C., & Specker, S. (2009). Increased mortality in bulimia nervosa and other eating disorders. *American Journal of Psychiatry, 16,* 173–176.

Diagnostic and Statistical Manual of Mental Disorders. 5th edn (DSM-5). (2013). American Psychiatric Association, Washington, DC.

Gary, A., Campbell-Ruggaard, J., Goodheart, K. L., & Clopton, J. R. (2012). The physiology of anorexia nervosa. In K. L. Goodheart, J. R. Clopton, & J. J. Robert-McComb (Eds.), *Eating disorders in women and children* (pp. 47–59). Boca Raton: CRC Press.

Gleaves, D. H., & Latner, J. D. (2008). Evidence-based therapies for children and adolescents with eating disorders. In R. G. Steele, T. D. Elkin, & M. C. Roberts (Eds.), *Handbook of evidence-based therapies for children and adolescents: Bridging science and practice* (pp. 335–353). New York: Springer.

Gleaves, D. H., Miller, K. J., Williams, T. L., & Summers, S. A. (2000). Eating disorders: An overview. In K. J. Miller & J. S. Mizes (Eds.), *Comparative treatments for eating disorders*. New York: Springer.

Goldsmith, T. (2006). *Bulimia: Binging and purging*. Physch. Central. Available at: http://psychcentral.com/lib/bulimia-binging-and-purging/000283. Accessed 25 Mar 2015.

Gore, S. A., Vander Wal, J. S., & Thelen, M. H. (2001). Treatment of eating disorders in children and adolescents. In J. K. Thompson & L. Smolak (Eds.), *Body image, eating disorders, and obesity in youth: Assessment, prevention, and treatment* (pp. 293–311). Washington, DC: American Psychological Association.

Grant, J. E., Kim, S. W., & Eckert, E. D. (2002). Body dysmorphic disorder in patients with anorexia nervosa: Prevalence, clinical features, and delusionality of body image. *International Journal of Eating Disorders, 32*, 291–300.

Hebebrand, J., Exner, C., Hebebrand, K., Hotkamp, C., & Casper, H. (2003). Hyperactivity in patients with anorexia nervosa and in semi starved rats: Evidence for a pivotal role in hypoleptinemia. *Physiology & Behavior, 79*, 25–37.

Hill, K., & Pomeroy, C. (2001). Assessment of physical status of children and adolescents with eating disorders and obesity. In J. K. Thompson & L. Smolak (Eds.), *Body image, eating disorders and obesity in youth: Assessment, prevention, and treatment* (pp. 171–191). Washington, DC: American Psychological Association.

Hoek, H. W. (2006). Incidence, prevalence and mortality of anorexia nervosa and other eating disorders. *Current Opinion in Psychiatry, 19*, 389–394.

Hoes, M. L., & Curtis, B. (2012). Pharmaceutical approaches for treating eating disorders. In K. L. Goodheart, J. R. Clopton, & J. J. Robert-McComb (Eds.), *Eating disorders in women and children* (pp. 415–429). Boca Raton: CRC Press.

Horowitz, L. M. (2004). *Interpersonal foundations of psychopathology*. Washington, DC: American Psychological Association.

Kaplan, A. S., & Howlett, A. (2010). Pharmacotherapy for anorexia nervosa. In C. M. Grilo & J. E. Mitchell (Eds.), *The treatment of eating disorders: A clinical handbook* (pp. 175–186). New York: Guilford Press.

Kluck, A. S., Garos, S., & Johnson, L. (2012). Sexuality and eating disorders. In K. L. Goodheart, J. R. Clopton, & J. J. Roberet-McComb (Eds.), *Eating disorders in women and children* (pp. 181–194). Boca Raton: CRC Press.

Mitchell, J. E., & Crow, S. J. (2010). In W. S. Agras (Ed.), *Medical comorbidities of eating disorders* (pp. 259–266). New York: The Oxford University Press.

Pike, K. M., Loeb, K., & Vitousek, K. (2001). Cognitive-behavioral therapy for anorexia nervosa and bulimia nervosa. In J. K. Thompson (Ed.), *Body image, eating disorders, and obesity: An integrative guide for assessment and treatment* (pp. 253–302). Washington, DC: American Psychological Association.

Pomeroy, C. (1996). Anorexia nervosa, bulimia nervosa, and binge eating disorder assessment of physical status. In J. K. Thompson (Ed.), *Body image, eating disorders, and obesity* (pp. 177–203). Washington, DC: American Psychological Association.

Robert-McComb, J. J., & McCullough, B. (2012). The physiology of bulimia nervosa. In K. L. Goodheart, J. L. Clopton, & J. J. Robert-McComb (Eds.), *Eating disorders in women and children* (pp. 61–74). Boca Raton: CRC Press.

Romano, S., Halmi, K., Sarkar, N., Koke, S., & Lee, J. (2002). A placebo-controlled study of fluoxetine in continued treatment of bulimia nervosa after successful acute fluoxetine treatment. *American Journal of Psychiatry. 74, 159*, 96–102.

Santuccis, P. (2014). *Glossary of treatment terms. A brief overview of therapies used in the treatment of eating disorders; A consumer's guide*. National Association of Anorexia Nervosa and

Associated Disorders. Available at: http://www.anad.org/get-information/information-about-treatment. Accessed 25 Mar 2015.

Simic, M., & Eisler, I. (2012). Family and multifamily therapy. In J. Fox & K. Goss (Eds.), *Eating and its disorders* (pp. 260–279). Malden: Wiley-Blackwell.

Stice, E., Wonderlich, S., & Wade, E. (2006). Eating disorders. In M. Hersen, J. C. Thomas, & R. T. Ammerman (Eds.), *Comprehensive handbook of personality and psychopathology* (pp. 330–347). Hoboken: Wiley.

Treasure, J. (2004). *A guide to the medical risk assessment for eating disorders*. London: Maudsley Publications.

Wassenaar, D., le Grange, D., Winship, J., & Lachenicht, L. (2000). The prevalence of eating disorder pathology in a cross-ethnic population of female students in South Africa. *European Eating Disorders Review, 8*, 225–236.

Watson, T. L., & Andersen, A. E. (2003). A critical examination of the amenorrhea and weight criteria for diagnosing anorexia nervosa. *Acta Psychiatrica Scandinavia, 108*, 175–182.

Wilson, G. T., Fairburn, C. G., Agras, W. S., Walsh, B. T., & Kraemer, H. (2002). Cognitive-behavior therapy for bulimia nervosa. *Journal of Consulting and Clinical Psychology, 70*, 267–274.

Woolsey, M. M. (2002). *Eating disorders: A clinical guide to counseling and treatment*. Chicago: American Dietetic Association.

Worobey, J., & Schoenfeld, D. (1999). Eating disordered behavior in dietetics students and students in other majors. *Journal of the American Dietetic Association, 99*, 1100–1102.

Chapter 13
Overweight and Obesity: The Bane of Modern Times

Abstract The chapter discusses, at length, overweight and obesity, the bane of modern times, a historical perspective, their scientific definition, their psychological fallout, their biological impact, and the interrelationship between energy intake and behavioral changes. There would also be a discussion on "very-low-calorie diets" and what role do they play on weight loss. Discussion would also be there on diet composition and how to manage over weight.

Keywords Overweight · Obesity · Psychological fallout · Energy balance · Behavioral change · "Very-low-calorie diet"

Compared to the eating disorders, discussed in the previous chapter, overweight and obesity have turned out to be a global pandemic (Popkin et al. 2012). It is a paradox that in a world bedeviled by food scarcity and starvation, even in developed countries, a sizeable population in the world should suffer from overweight and obesity, leading these symptoms to health problems and psychological distress. Excess body weight leads to several health problems including adverse social and psychological consequences. Because of the physiological and psychological consequences of overweight and obesity, weight loss has become a preoccupation for many people and an industry in itself. Commercial interests are trying their best to grab the pecuniary opportunity. This preoccupation is evident in TV shows, in supermarkets, in bookstores, in the physician's office, and in the ever-growing popularity of diet centers and fitness clubs. Although some individuals manage to lose weight, maintaining low body weight for a considerable length of time becomes very difficult. Yet, the sheer number of individuals who are at risk of obesity and its consequent complications demands something be done to stem the tide of what appears to be an overweight epidemic. USA is one country in the world where it has reached an epidemic proportion. Recall the striking increase in research in eating disorders mentioned in the previous chapter. The increase in research on obesity has been even more dramatic, and, arguably, more urgent. To illustrate the point, the MEDLINE citations for "obesity" in the USA in 1993, the inaugural year of the journal *Obesity*, totaled 1332. In 2013, that

© Springer Nature Switzerland AG 2020
K. P. Nair, *Food and Human Responses*,
https://doi.org/10.1007/978-3-030-35437-4_13

number jumped more than 500–9079%. It is, then, clear that physicians, psychologists, public health workers, and politicians consider obesity to be *the public health problem of the* twenty-first *century*.

13.1   Obesity and Overweight: How Do We Define Them?

The Latin term *obesus* means to devour and suggests that overeating was recognized early on as a contributing factor to the condition we call obesity. While in bulimia, binge eating may characterize some individuals who are overweight, we now recognize that obesity results from a myriad of causes, not simply due to overeating, although food intake, of course, plays a substantive role.

13.2   Obesity: A Historical Perspective

Though obesity is a modern widespread problem, relics unearthed by archaeologists indicate that overweight women were not only observed during the Stone Age, but, in some prehistoric cultures, it was even revered. Small statuettes of women with pronounced abdominal obesity and pendulous breasts have been found across all of Europe and may date back to 25000 years to the Paleolithic era. Numerous figurines excavated in Turkey, purported to be over 7500 years old, from the Neolithic and Chalcolithic periods, are said to represent the mother goddess and display the exaggerated hips, bellies, and breasts that may also characterize fertility (Bray 2014). In many African cultures, even now, women with generous hips are preferred than lean and thin ones. Thus, the ancient runs into the present. Excessive overweight was not unknown to the Egyptians of 2000 BCE, as the mummies of stout kings and queens will attest. Although royal mummies represented the higher classes, even then obesity was regarded as objectionable (Darby et al. 1977). The clinical treatment of obesity was first addressed by Hippocrates in the fifth century, BCE, who suggested hard work before eating, if one desired to lose weight, and a limit of one meal, a day. Nearly 2000 years ago, Galen declared that running, along with food which provided little nourishment, made his stout patients moderately thin in short order. To these Greco-Roman roots, we can add acupuncture by the ancient Chinese, medicinal therapy by the Indians, like *Ayurveda*, and diet therapy by the Arabs, as early approaches to the treatment of obesity. Based on this evidence, it is abundantly clear that obesity, and the role of diet and exercise in treating obesity, were recognized thousands of years ago.

13.3   How to Quantify Obesity and Overweight?

Unlike the current eating disorders, obesity is not a psychological disturbance, though there is a psychological correlation, rather, is a state of disease characterized by the excessive accumulation of body fat (Aronne et al. 2009). Stated simply, the physical measurement of body fat or its approximation defines the state of being overweight or obese. There are numerous techniques now available to measure body fat directly or indirectly. Direct measures include underwater weighing, total body water analysis, total body K analysis, bioelectrical impedance analysis, ultrasound (ultrasonography, USG), computerized tomography (CT scan), magnetic resonance imaging (MRI), dual energy X-ray absorptiometry, and neutron inelastic scattering (Goodpaster 2002). Although this wide array of techniques indicates that measuring body fat is now a routine matter, and can be done with extreme accuracy, none of these approaches are ideal. Space constraints, measurement error, difficulty in interpretation, radiation exposure, and the lack of suitability for the severely obese limit many of these procedures. Furthermore, the specialized equipment, many of these methods require, makes them extremely expensive and inconvenient for large studies (DiGirolamo et al. 2000) and may be out of reach for many poor patients. On account of the above described drawbacks, indirect measures to assess body weight are more frequent in clinical settings. Because the skin is anywhere from 0.5 mm to 2 mm in thickness, with subcutaneous fat just below the surface of the skin, *skinfold thickness* can be determined across multiple sites, such as the mid-triceps, shoulder blade, and abdomen (Garcia et al. 2005). Special calipers are used to measure the skinfolds, and substantial skill and training are required to follow the procedure. Although there is some disagreement about the sites which most accurately reflect actual body fat, its utility with the elderly who accumulate fat in additional places, and its tendency to underestimate fat in cases of greater obesity, measuring skinfold thickness is nonetheless a useful and more economical approach to estimate body fat. Waist and hip circumference, with a computed *waist-to-hip ratio*, has also been used to approximate visceral adipose tissue. Although they may be somewhat inaccurate as an actual measure of body fat, they are suitable to distinguish between categories of percentage body fat (Flegal et al. 2009).

In this connection, mention must be made of the Egyptian patient, Eman, who came to Mumbai (India) for weight loss surgery, known as bariatric surgery, which is done on 200,000 patients annually, the world over, with a success rate of about 83% in diabetics and 60% average weight loss in 5 years among other overweight patients. Eman could shed close to 250 kg in about 3 months' time, and she went back to Egypt weighing much less. She was in excess of 500 kg when she came to India for the surgery. The successful surgery made headline news all over the world, in particular, in India.

The long used Quetelet Index (Quetelet 1871), or *body mass index* (*BMI*), as it is more commonly known, has been endorsed by the National Institutes of Health of the USA, the National Audit Office of England, and the International Obesity Task Force of the World Health Organization (WHO), as providing the most inexpensive, yet clinically relevant, assessment of obesity across varied populations. BMI describes relative weight for height and is calculated as follows:

$$BMI = \frac{\text{Weight in Kilograms}}{\text{Height in Meters}}$$

or

$$\frac{\text{Weight in Pounds} \times 704.5}{\text{Height in Inches}}$$

BMI classifications of different categories of individuals are shown in the following table (Table 13.1).

BMI is significantly correlated with total body fat content, a more precise measure of total body fat than weight alone, and is a better estimate of health risk than are any actuarial tables from insurance companies. Despite the widespread acceptance of BMI as the most practical measure of overweight and obesity in adults, it must be remembered that weight and height are only a proxy for body composition. For example, a body builder may be classified as "overweight" by BMI standards, but it would derive from muscle mass and not excess fat. Additionally, BMI is less accurate with individuals who are extremely short or tall and cannot gauge whether someone who is of normal weight, but is extremely sedentary, might have excess body fat. For these reasons, clinical assessments are imperative to interpret BMI in situations where body weight may be affected by increased muscularity or by disease.

Table 13.1 Body mass index (BMI) classification for different individuals

Individual's BMI (kg/m²)	Obesity class weight	Status
Underweight	<18.5	
Normal	18.5–24.9	
Overweight	25.0–29.9	
Obese	30.0–34.9	I
Moderately obese	35.0–39.9	II
Extremely obese	40.0–49.9	III
Super obese	>50.0	IV

Source: Sturm (2007)

13.4   Prevalence of Overweight and Obesity

During the last three decades, prevalence of obesity has increased at an alarming rate – not just in developed countries but all over the world. In fact overweight has begun to displace under nutrition and infectious diseases as the most common health problems. Between 1980 and 2013, overweight and obesity increased by 27.5% for adults. For children, it is even worse at 47.1% during the same period. While the prevalence of obesity tends to be lower in less developed countries than in affluent ones, obesity rates have risen in lower- and middle-income countries (Jain 2004). Nations which had formerly faced significant problems of malnutrition, and famine, have now attained adequate food supplies – classically represented by India – accompanying changes in diet composition. At the same time, their occupational and other types of physical activity have decreased, which is another contributing factor to overweight and obesity. Worldwide, it is estimated that the number of individuals who are overweight and obese increased from 857 million in 1980 to 2.1 billion in 2013 (Ng et al. 2014). It is significant to note that of the 671 million obese individuals in the world, over half of them live in ten countries, which, in order of their numbers, are the USA, China, India, Russia, Brazil, Mexico, Egypt, Germany, Pakistan, and Indonesia.

13.5   Overweight and Obesity and Its Sociocultural Correlates

Increased body weight is related to many sociocultural aspects. The following table (Table 13.2) summarizes some of these.

The sociocultural parameters can be classified as socio-personal, social context, and socioeconomic status. Socio-personal characteristics are inherent in the person and include, age, sex, and race/ethnicity. Both men and women appear to gain the most weight between 25 and 34 years of age. Thereafter, weight may continue to increase, but at a slower rate, and show a decline after 60 (Elia 2001). Interestingly enough, men are likely to be overweight than women, but women are likelier to be obese than men, a pattern evident throughout much of the world (Ng et al. 2014). In the USA, black and Latina women have higher rates of overweight individuals than non-Hispanic white women, followed by women of Asian heritage (Wang and Beydoun 2007).

13.6   The Physiological Fallout of Overweight and Obesity

It is not the question of physical appearance that is central to overweight and obesity. The associated risks for illness and disease are what make this a significant public health problem. The following table (Table 13.3) summarizes this aspect.

Table 13.2 Socioeconomic correlates of obesity

Parameter	Low-/middle-income countries	High-income countries
Education	+	–
Employment	~	–
Income	+	–
Occupation	–	–
Material possessions	+	–

Source: McLaren (2007)
Note:
+ = A positive correlation
– = A negative correlation
~ = An uncertain association

Table 13.3 Obesity and comorbidities

Physical Risks: Type 2 diabetes mellitus, congestive heart failure, stroke, hypertension, atrial fibrillation, malignancies, obstructive sleep apnea, osteoarthritis, gout, gallbladder disease, kidney stones, renal failure, and reproductive complications
Psychosocial Risks: Depression about weight and eating, poor body image, societal discrimination, and impairments in executive functioning

Source: Pedersen et al. (2012)

Heart failure, chief cause for death in the USA, is associated with a BMI of >30. Also hypertension, stroke, diabetes, gallbladder, joint diseases, and some forms of cancer, as well, are associated with overweight and obesity. Overweight at 40 is associated with a 3-year decrease in life expectancy, and obesity decreases life expectancy by 6–7 years. Hypertension is more common in the overweight individuals than others who are not. In combination with high blood cholesterol and serum triglycerides, both linked to overweight and obesity, fatality is a definite risk.

Of all the health risks, adult onset of diabetes (Type 2) appears to have the strongest association with overweight and obesity. About 85% of diabetics can be classified as Type II, and, of these, approximately 90% are obese. Obesity compromises glucose tolerance and increases insulin resistance, and diabetes, in turn, may cause heart and kidney diseases and vascular problems. Interestingly enough, the location of body fat seems to be a risk factor for disease, independent of the fat itself. The "apple" shape, where fat is collected in the abdominal area, is more characteristic of men, who are obese, hence the name *android* (man-like). By contrast, women typically collect fat in their hips and buttocks, giving them a "pear" shape known as gynecoid (woman-like). See the illustrative figure below (Fig. 13.1).

In general, the "apple" shape in men is associated with greater health risks than the "pear"-shaped pattern more often seen in women, since excess fat in the abdominal region appears to raise blood lipid levels, which then interfere with insulin function (Despres and Lemieux 2006). Visceral fat (the fat which surrounds the organs of the abdominal cavity) is linked to hyperlipidemia, hypertension, heart disease, and diabetes. It is also observed that several types of cancer are associated

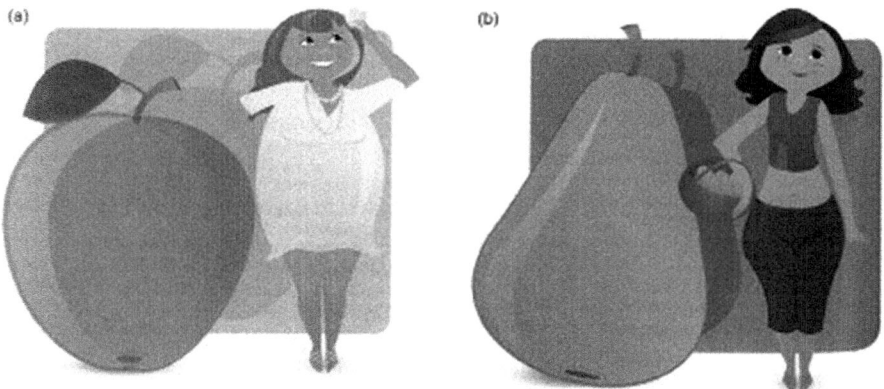

Fig. 13.1   (**a**) The apple and (**b**) pear shapes which characterize obesity. (*Source:* Shutterstock)

with overweight and obesity, especially with excess abdominal fat. Overweight men are at a great risk of developing prostate, colon, and rectal cancer. Overweight women are at a greater risk of developing cancer of the colon, gall bladder, uterus, cervix, ovary, and breasts. At present, however, it is not clear whether the incidence of cancer is associated with excess weight or a high-fat, high-calorie diet (Roberts et al. 2010). Excess weight which accompanies obesity also appears to increase the risk of developing osteoarthritis, gout, gall bladder diseases, and obstructive sleep apnea.

13.7   The Psychological Fallout of Overweight and Obesity

Contrary to some wrongly held view, neither overweight nor obesity is a psychological disturbance, and the rates of general psychopathology do not appear to be greater in these groups, compared to the normal weighted or nonobese individuals. Moreover, there is but little evidence that these individuals differ from the nonobese on general personality or temperament traits, such as extraversion/introversion, rigidity/flexibility, assertiveness, locus of control, and emotionality (Worobey 2002). On the other hand, obese individuals who seek treatment for their obesity often display a higher degree of anxiety and depression than does the general population, though, this is also true for many individuals who seek treatment for medical or surgical reasons (Tuthill et al. 2006). All things considered, there is but little evidence for an "obese personality." Being distressed enough to seek treatment may be the critical variable, as the obese who seek treatment report more symptoms of personality disorder than comparable obese individuals who do not (McElroy et al. 2004).

As is true of adults, obesity and psychological problems in childhood appear to be bidirectional, as psychological distress may lead to weight gain, but obesity can also lead to psychological problems (Puder and Munsch 2010). The most frequent psychological problems faced by obese children can be categorized as "externalizing behavior," such as impulsivity, and internatilizing behavior, such as depression.

Body image disparagement and poor self-esteem are the likely result of the tremendous disdain which western societies seem to have for the obese. In the US society, for instance, overweight and obesity are socially stigmatized, and the perception of others is that the obese individual has but little self-control. Unlike a physical deformity or the skin color, obesity is viewed as intentional, with the obese individual held accountable and responsible for his or her condition (Brownell et al. 2010). In societies, where self-worth is tied to physical appearance, an individual's body is always on display, and, if that body is fat, the individual is discriminated against, most unfortunately. In a recent study, individuals who reported experiencing under treatment, due to their overweight, had 2.5 times higher risk of becoming obese by the end of the 4-year follow-up (Sutin and Terracciano 2013). There are various kinds of societal biases against the obese, like not letting out apartments to live, gaining college admissions or, even in limiting a job. Negative attitude toward the obese has even been reported by parents, in providing comparatively less financial assistance for their overweight daughters, and among health professionals, including obesity specialists (Schwartz et al. 2003). Recent research has added cognitive performance to the list of consequences on which overweight and obesity may add their influence. Given all the above-discussed correlates, it is obvious that the quality of life for the obese and overweight individuals, from both a physiological and psychological level, can be markedly compromised. In a landmark investigation of the self-perceived impact that obesity had on their lives, Rand and MacGregor (1991) interviewed patients who had lost 100 lbs (45 kg) or more through gastric bypass surgery. The patients indicated that they would rather be blind, deaf, dyslexic, diabetic, or even an amputee rather than return to their earlier morbidity obese status. Such a strong attitude toward their formerly obese state suggests that life was fairly intolerable for these individuals, with the discomfort, prejudice, ridicule, and self-loathing which they experienced far exceeding any enjoyment they may have derived from eating. A recent investigation suggests that becoming an overweight or obese individual is not just a matter of excess food intake, but a combination of both psychological and behavioral aberrations.

13.8   Obesity: Its Etiology

It is important to realize that while the earlier discussion concentrated on the behavioral aspects of obesity and its psychological correlates, the etiology of obesity and overweight has not been addressed. The basic criterion for gaining excess weight or becoming an obese is that *over an extended period of time, an individual's energy intake is greater than his or her energy expenditure.* When energy expenditure is lower than energy creation through food intake, a positive balance occurs, leading to weight gain in the individual's body. Conversely, a negative energy balance results in a decrease in fat stores, with subsequent weight loss (Das and Roberts 2012). Body weight is regulated through a series of physiological processes which have the capacity to maintain a stable weight within a narrow range, or set point, by virtue of a feedback control mechanism.

As it is now recognized that obesity has many interrelated causes, a discussion on the current explanations for variations in weight gain which lead to overweight or obesity will form the principal discussion in the following sections.

13.9   The Biological Influences of Obesity and Overweight

Just about 10% of the population alone was classified as obese prior to the twentieth century. And this was only for genetic reasons (Helmchen and Henderson 2004). Over the last one century, the population's gene pool has hardly changed, but there is an exponential increase in obesity during the last three decades. For most individuals, then, gene is not the culprit in causing obesity, though, admittedly, it may influence certain processes that help lead to obesity. Nevertheless, there is an emerging view that heredity has a part to play in obesity. For instance, 48% children with obese parents became overweight by the age between 9 and 10, compared to 13% with normal weight parents. Thus, children seem to share the eating environment with their parents, as well as their genes. It is also important to examine the concordance rate, or similarity, of obesity between twins. The Stunkard research group has carried out some remarkable research in this area using adopted children, showing that the BMI of biological parents corresponds highly to their children given for adoption. In turn, no correspondence was shown between the adoptive parents and the children they adopted. Moreover, in examining twins who were reared together versus apart, these investigators found that the correlation in weight between identical twins reared apart was about twice as high as for the fraternal twins who were reared together. Research now suggests that a large number of genes, in fact, may have a crucial role to play in determining an individual's obesity. For example, an obesity gene called "ob" has been identified that is expressed in the fat cells and codes for the protein leptin – the word is derived from Greek *leptos*, meaning thin. Mice with defective *ob* genes do not produce leptin, and they become obese. However, they would lose fat and subsequently become thin if injected with leptin (Ahima et al. 2000). But, obese individuals generally have high levels of leptin, and leptin concentration generally increases with weight gain. It is, therefore, suggested that leptin level rises with weight gain in order to suppress appetite, but the obese are resistant to its action in the way that individuals with Type 2 diabetes have high levels of insulin but are resistant to its action insulin resistance seen among Type 2 diabetics.

In contrast to leptin, which acts as a hormone in the hypothalamus, ghrelin is a peptide which is produced in the gastrointestinal tract which functions as a neuropeptide in the CNS. Known as the "hunger hormone," ghrelin plays a significant role in regulating hunger as well as in the distribution and rate of use of energy

(Burger and Berner 2014). In rodent models, administering ghrelin stimulates food intake and increases body weight, while in human beings, leptin levels are higher in obese individuals than in lean individuals as indicated earlier. Circulating ghrelin levels appear to be negatively correlated with BMI, which suggests that it may operate as a peripheral signal for energy insufficiency. Indeed, as based on their reciprocal fluctuation, ghrelin and leptin have been proposed as comprising a yin-yang sort of regulatory system which allows for the precise tuning of energy balance and ultimately body weight (Fernandez-Fernandez and Tena-Sempere 2013). Genes which code for *uncoupling proteins* may also prove to be important in understanding the development of obesity (Miyashita and Hosokawa 2013). The human body has two types of adipose tissues, the mostly white adipose tissue which stores fat for other cells to use later as energy and brown adipose tissue which releases stored energy as heat. When white adipose tissue is oxidized, the most of the energy is captured in adenosine triphosphate (ATP), and the remainder is released as heat. In the brown adipose tissue, oxidation may be uncoupled from ATP formation producing only heat. This results in the body spending energy instead of storing it. The gene which codes for uncoupling protein 2 is active in both brown and white tissue fat and also influences BMR (Fleury et al. 1997). Animals with high amounts of this protein appear to resist weight gain, while those with minimal amounts gain weight rapidly (Morton et al. 2004). Last, but not the least, the development and metabolism of fat cells themselves may also contribute to overweight and obesity. The amount of fat in a person's body reflects both the number and size of the fat cells that he or she possesses. The *number* of fat cells increases most rapidly during late childhood and early adolescence and more rapidly in obese children than in the lean ones. As fat cells fill with fat droplets, they increase in size and may also divide if they reach their maximum number. Obesity can thus develop when a person's fat cells increase in number, size, or both. With overfeeding of adult individuals, increasing fat cell size correlated with upper body fat gain, but the number of fat cells was not significantly changed. However, while there was no change in the size of the lower body fat cells, during the experiment, the number of lower body fat cells did significantly increase.

13.10   The Interrelationship Between Energy Intake and Behavioral Changes

Given what we currently know about the multiple causes of obesity, the clue does not entirely lie in energy intake, because despite higher energy intake, some individuals remain thin while others put on fat. For instance, diet histories of obese individuals often indicate energy intakes and patterns similar to those of normal weight individuals. However, we must admit that diet histories are not the most accurate records of actual intake, as both obese and nonobese people commonly underestimate their dietary intake (Johansson et al. 2001). Beyond total energy intake, obese individuals tend to eat more fat, and it is quite clear that a high-fat diet also promotes obesity, especially in those individuals who are genetically predisposed to obesity. Relative to protein and carbohydrate, dietary fat provides more kilocalories per gram and requires less energy to be metabolized, which further amplifies energy intake and increases body fat stores. Additionally, high-fat foods are highly preferred by most individuals; therefore, more such foods are likely to be eaten. Some have argued that foods higher in fat content have lower satiating power relative to protein and carbohydrates (Paddon-Jones et al. 2008), which can, therefore, lead to passive overconsumption (Blundell and MacDiarmid 1997).

Quite recently, there has been much attention on the hypothesis that it is not fat; instead, it is carbohydrates, in particular, refined carbohydrates (e.g., high-fructose corn syrup), which are more likely to cause obesity (Bray 2008). However, there are detractors, among the scientific community, to this line of thinking (Nestle and Nesheim 2012), although most experts agree that an excess intake of sugar-sweetened beverages by the American public is a significant contributor to weight gain, in particular, during childhood (Kavey 2010).

Anyway, one thing is absolutely clear. The food habits of the western civilizations show that people there are definitely consuming more calories than what they require. In the USA, for example, food consumption has jumped 21.7% during the last four decades, an increase of 459 calories per day (Economic Research Service 2014). Wherever fast food restaurants now proliferate, the competition between corporate chains has resulted in promotions which appeal to customers via the upgrading of standard meals – a practice known as "super sizing" (Rolls 2003). To a meal, which is already high in fat, sugar, and Na, the customer is able to further enlarge it with a soft drink (carbonated water which is sweet adding to the calories) or French fries, for a minimal increase in price. Additionally, these fast food chains are supplemented by cafeteria-style restaurants , which offer "all you can eat" buffets with a tremendous variety of foods for one low price, all contributing to what is now referred to as our "obesogenic" environment (Kirk et al. 2010). The dietary pattern of both humans and animals indicates that food consumption increased directly proportional the varieties at disposal, which lead to increase body fat and obesity. Self-control is thrown to the winds and binge eating follows. With children, when parents use threats or rewards, to coerce their children to eat all that is on the plate, seem to inadvertently interfere with the child's ability to self-regulate energy intake, by forcing them to ignore internal feelings of satiety. Conversely, being overly restrictive of certain desirable foods may also have unintended consequences. The Birch research group has conducted exemplary experiments on the phenomenon of children's "eating in the absence of hunger" and have established that if mothers overly restrict tasty snacks at home, their daughters are likelier to eat such snacks given the opportunity even after indicating that they were not hungry. Since most children will eat the so-called junk foods, if they are available in the house, some have further suggested that an indulgent approach toward their children's eating habits may also be inadvisable, as it has been implicated in children's gaining excess weight. It is thus apparent that a complex relationship exists between parental control and children's ability to self-regulate their energy intake. More research, clearly, in this area is warranted.

13.11   The Interrelationship Between Energy Expenditure and Behavioral Changes

Given the dramatic increase in obesity during the last two decades, there is a case for suggesting that the increased prevalence of overweight and obesity is due to lack of physical activity among adults and children. The current sedentary life style, thanks to technological innovations, has contributed to this malaise. The following table (Table 13.4) gives an idea of energy expenditure stemming from different activities.

If physical inactivity is deemed an important contributor to obesity, television watching contributes the most to underactivity (Hu et al. 2003). Television watching is, in principle, a sedentary activity and worse, people who watch the TV, munch on snacks which are invariably food items that contribute to fat build up. Most disturbingly, many high-calorie/high-fast foods are advertised on TV, which will, undoubtedly, influence the watcher to purchase these junk foods and consume – a health hazard. The effect is most widely seen among children.

Children now ride buses to school instead of walking, as was the practice, decades earlier, and there is much less time spent on physical education in school. When they back home after the classes, they are often confined indoors, which lead them to watch TV, and the negative fallout takes a toll on their weight and health.

Both physical activity and exercise are the two major components of energy output, but recent research has identified another form of energy expenditure which may prove to be of some importance in explaining why certain individuals are more resistant to weight gain than others. The energy we use even when we are not exercising or attempting to be active, rather the energy that is associated with fidgeting, crossing, or uncrossing the legs, and otherwise maintaining one's posture, is known as non-exercise activity thermogenesis (NEAT, Levine 2004). In an investigation, where the experimental subjects were overfed, and, after accounting for BMR, dietary-induced thermogenesis, and physical activity, those who were higher in NEAT were most resistant to weight gain (Levine et al. 1999). Further studies of this newly identified form of energy expenditure would seem to be warranted (Levine et al. 2006).

Table 13.4 Estimates of energy expenditure for different activities

Type of activity	kcal/h/kg	kcal/h/lb	50 kg/110 lb	68 kg/150 lb	91 kg/200 lb
Aerobics					
Light	3.0	1.36	150	205	273
Medium	5.0	2.27	250	341	455
Heavy	8.0	3.64	400	545	727
Walking					
Strolling <2 mph	2.0	0.91	100	136	182
Brisk pace, level	4.0	1.82	200	273	364
Moderate pace					
Uphill	6.0	2.73	300	409	545
Running					
Jogging	7.0	3.18	350	477	636
Running 7 mph	11.5	5.23	575	784	1045
Running 10 mph	16.0	7.27	800	1091	1455
Bicycling					
Leisurely <10 mph	4.0	1.82	200	273	364
Moderate 12–14 mph	8.0	3.64	400	545	727
Racing 16–19 mph	12.0	5.45	600	818	1091
Recreation					
Hay sack	4.0	1.82	200	273	364
Golf	4.5	2.05	225	307	409
Rollerblading	7.0	3.18	350	477	636
Sports					
Ultimate frisbee	3.5	1.59	175	239	318
Downhill skiing	6.0	2.73	300	409	545
Singles tennis	8.0	3.64	400	545	727
Daily activities					
Sleeping	1.2	0.55	60	82	109
Studying/writing	1.8	0.82	90	123	164
Cooking	2.5	1.14	125	170	227
Household upkeep					
House painting	4.0	1.82	200	273	364
Gardening	5.0	2.27	250	341	455
Shoveling snow	6.0	2.73	300	409	545

Source: Nieman (2010)

13.12   The Interrelationship Between Sleep and Energy Balance

There is now a strong body of evidence, gathered over the past decade or more, to suggest the positive correlation between shorter duration of sleep and obesity among both adults and children. A meta-analysis spanning many countries, worldwide, which represents 36 different population samples and 634,511 subjects, shows that the risk of obesity is 50% or more among adults getting less than 5 h sleep and among children getting less than 10 h sleep (Cappuccio et al. 2008). Current research focuses on the subject how sleep may inverse the risk of obesity. One theory suggests that sleep deprivation seems to result in lowered levels of leptin circulation and increased levels of ghrelin, which, as discussed earlier, enhances appetite. Evidence also exists that shortened sleep results in abnormalities in glucose metabolism which can lead to insulin resistance and subsequently greater fat storage (Lucassen et al. 2012). Moreover, it has also been noted that shortened sleep can enhance concentration of leukocytes and cytokines which may promote an inflammatory state such as that is associated with Type 2 diabetes (Knutson et al. 2007). While it is obviously very clear that shortened sleep and the development of obesity are linked, attributing causality to shortened sleep will depend on further research, as it is also possible that obesity itself may lead to disturbances in sleep (Miller and Cappuccio 2013).

13.13   How to Lose Weight?

In the world, it is the USA which spends an enormous amount of money, well over US $ 60 billion/year on products and efforts aimed at helping people lose weight (Marketdata 2014). However, many individuals who wish to shed excess weight do not need to do so, many who need to shed excess weight are unsuccessful, and of the few who succeed, even fewer meet with permanent success. The last five decades have seen an evolution in our views of obesity, as we have moved from a belief that failed will power leads to obesity which must be corrected to a recognition that obesity results from multiple factors, but physical health is more important than weight loss. As early as the 1940s, the self-help group model first appeared, with commercial programs which sold pre-packaged low-calorie meals emerging in the 1950s and 1960s. In the 1970s, behavioral approaches which focused on changing eating habits were implemented. The value of aerobic exercise was endorsed next, with the slogan "no pain, no gain," a popular mantra. Throughout the 1980s, the use of surgical treatments was introduced, and in the 1990s, the belief that drugs could serve as a "magic bullet" was formally tested. Over the last decade, we have seen tremendous growth in the sale of herbal formulations, which has been discussed earlier in this book, with many being used in the hope of losing weight. Given the scope of the obesity problem, all of these approaches are still employed in one form or another. As the focus of this book is on nutrition, vis-à-vis food, the discussion will restrict to approaches which rely on the individual as actor, as opposed to a physician/surgeon (as in the case of bariatric surgery) or a pharmacist (as in the case of prescription drugs). The reader, however, is recommended to refer some excellent reviews which cover the surgical approaches (Manco 2013) and weight loss drugs (Schwartz and Savastano 2013). The word "diet" has its Greek origin, from the word *diaita*, which means "manner of living," which is derived from the root word *diaitasthai*, which means "to lead one's life." For the majority of overweight and obese individuals, dieting is a way of life and represents their primary strategy to lose weight.

13.14   What Are The "Very-Low-Calorie Diets," and What Role Do They Play in Weight Loss?

In the 1980s, centers which catered to obese individuals for their rehabilitation began to offer what was then known as "very-low-calorie diets" (VLCDs), which is an approach deemed suitable for individuals who are moderately obese, that is, 41–100% of their ideal weight. These VLCDs had a calorie content around 1200–1500 calories/day, the typical VLCD plan would provide no more than 800 kilocalories, given via a liquid formula, or as lean meat, fish, or poultry. Clients of these weight loss centers would also be supplemented with vitamins and other mineral supplements. When carefully administered, under strict medical supervision, VLCDs appeared to be relatively safe, although side effects such as headache, dizziness, fatigue, dry mouth, cold intolerance, constipation, dry skin, menstrual irregularities, and hair loss have been reported (Saris 2001). While designed to be nutritionally adequate, the body still responds to this severe energy restriction as if the individual under treatment were starving, conserving energy, by reducing BMR and slowing fat oxidation. For this reason, a VLCD is only appropriate for short-term purposes, that is, 4–6 weeks, which may help the individual lose about 20 kg (44 lb) over the stipulated period, though much of it may be water and electrolytes (Pi-Sunyer 2007). Unfortunately, the near starving it creates primes the body to regain weight at the first opportunity, with two-thirds of the weight loss being regained within 1 year. Such a rapid loss of weight followed by a steady gain is likely to be detrimental to both physical and psychological health of the individual, and the use of VLCDs is now rarely employed.

13.15   The Role of "Yo-Yo Dieting" in Weight Loss

The tragedy of weight loss diets and/or exercises is that many individuals who try this will lose weight, initially, but, will not be able to maintain it over an extended period of time and will, inevitably, regain it over a period of time. Nevertheless, these individuals are, more often than not, likely to repeat this strategy, regain weight, and diet again, a behavioral pattern as weight cycling or "yo-yo dieting." An investigation with rats proved this pattern. Such a finding is in line with anecdotal reports by dieters that every attempt at dieting it takes longer to lose weight and quicker to gain it back. Recent research, albeit with a mouse model, seemed to demonstrate that food restriction altered stress and feeding pathways in the brain, which promoted binge eating of high-fat foods after the mice were exposed to stress.

13.16   How to Manage One's Weight?

In contrast to the extreme forms of dieting that have been discussed earlier, with the endless number of diet plans which are advertised, adopted, and subsequently abandoned, a more viable approach to treating obesity is one of striving for a healthy weight, rather than reducing one's weight to conform to a weight-for-height table discussed earlier. In fact, a consistent sub-theme within the Dietary Guidelines for Americans is to help Americans "achieve and maintain a healthy weight". Modest weight loss, even if a person remains overweight, can reduce the risk of heart attack and improve control of diabetes. A loss of 10–15 lb (4–5 kg), for instance, can lower an individual's BMI by two units, which will significantly improve his or her health. However, experts recommend a loss of no more than 1–2 lb/week, for a maximum of 10% of the total body weight per year. Not incidentally, the recommendation to "be physically active" has been in the Dietary Guidelines since the 1995 edition. Reducing blood pressure or cholesterol level through diet and exercise is therefore a more useful goal than a mere focus on weight, although the strategies used to do so will likely result in a healthier body weight and composition, as well. Ironically, the unrealistic expectations for weight loss that many people possess may prevent them from appreciating their actual success.

13.17   The Diet Composition

It will be more effective to target a small reduction (about 200–300 kilo calories/day) than attempting to subsist on a starvation diet of 1000–1200 kilocalories/day if one's aim is to reduce weight. A realistic energy intake should provide less energy than the person needs to maintain his/her body weight, but a daily intake of less than 1200 kilocalories could make it difficult to meet nutritional adequacy. Adequate intake will likelier ensure more successful weight loss than a severely restrictive plan that induces starvation and deprivation, which can lead to bingeing subsequently.

Restricting fat intake is an obvious strategy, and studies show that individuals who eat a low-fat, high-protein diet will more readily satisfy their hunger and eat less food (Noakes et al. 2005). As discussed earlier, such low-fat diets should not rely on excess carbohydrates or out-of-bounds portion sizes to compensate for the reduction in fat (Insel et al. 2013).

Factors related to obesity and its treatment are illustrated in the following figure (Fig. 13.2).

Even low-fat foods provide excessive calories if eaten in huge quantities, and excess refined carbohydrates (that is sugar) ought to be avoided (Johnson et al. 2009). Complex carbohydrate foods such as fresh fruits, vegetables, legumes, and whole grains are low in fats, but also rich in vitamins, minerals, and fiber. High-fiber foods are also beneficial because they require more effort to chew, in effect, slowing down intake, while having a strong satiety effect (Burton-Freeman 2000).

Water, that ubiquitous great blessing of nature, which assists the gastrointestinal tract in adapting to a high fiber diet, will also help to fill the stomach between meals and dilute the metabolic wastes which break down of fat generates. Adding water to a recipe to decrease energy density of the meal is, therefore, a useful strategy in maintaining weight (Rolls et al. 2005). It is also, perhaps, interesting to speculate that water which is extensively used in all the preparations of food in oriental tradi-

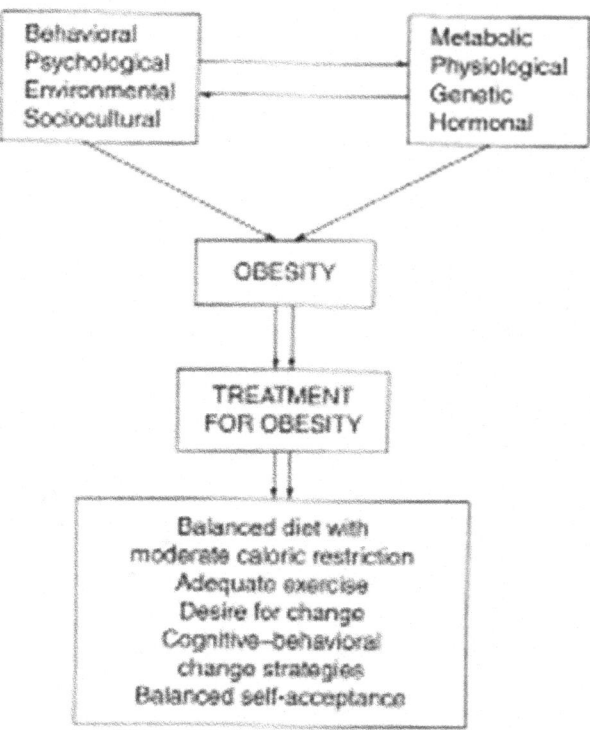

Fig. 13.2 Factors related to the obesity and its treatment. (*Source:* Insel et al. 2013)

tion, is an anthropological innovation, which has its beneficial fallout inasmuch as the body weight is concerned. And, it is not by accident that in the orient, people are, generally not overweight or obese, as compared to their compatriots in the west, who use much less water in preparing food, and end up being fatter eating such food. Adding more of just any liquid to one's diet is not as beneficial as adding plain water. While high in antioxidants, alcoholic beverages are also high in calories but low in major nutrients, as are regular soft drinks, with their high sugar content. At the same time, calories ingested as fluids are not compensated for, as well as solids – individuals do not eat any less when drinking such fluids, resulting in higher energy intake than if fruit, for example, were eaten plain (Mourao et al. 2007). And, the caffeine in coffee, which serves as a diuretic, gives only the illusion of weight loss.

13.18 The Role of Physical Activity in Weight Loss/Weight Control

Regular physical activity is an important component of the weight loss program. While aerobic exercises like physical walking, jogging, running, or thread mill exercise help burn the excess calories directly, anaerobic ones like weight lifting or sit-ups help build muscles. Additionally, regular exercise can discourage overeating by reducing stress, which also reinforce positive thinking and bring about an overall feeling of well-being and positive societal interaction. In this connection, mention must be made of yoga, a meditative posture combined with different bodily postures, which includes bodily exercise like *Surya Namaskar* (the word *Surya* in Sanskrit means sun, and the word *Namaskar* means worship, both combined meaning worshipping the sun"). This should be, in particular, done at the time the sun rises on the horizon. Overweight individuals who combine strict diet regime and exercise will more likely lose weight through loss of fat, retain more muscle, and regain less weight than those who only diet, over a period of time. A systematic review of 33 trials which evaluated the effectiveness of diet, exercise, or diet and exercise combined, concluded that diet with exercise produced 20% more weight loss than diet alone (Curioni and Lourenco 2005). Diet with exercise also resulted in a 20% greater weight loss after 1 year than did diet alone, although almost half of the initial weight loss was regained by the end of the year.

As has already been shown in the data in Table 13.4 (discussed earlier), physical activity has a direct effect on energy expenditure. While regular exercise of moderate intensity would improve health, to lose fat and weight, exercise should be as vigorous as physical shape and time will permit, with intensity offset by the amount of time at disposal. According to the Physical Activity Guidelines for Americans to merely maintain their weight, adolescents and children need to be physically active a minimum of 60 min a day. For adults, the equivalent of 150 min or more of moderate intensity aerobic activity each week is recommended. In order to lose weight, exercise levels should progressively be increased to 200–300 min (3.3–5 h/week). Activity also contributes to energy expenditure indirectly by speeding up the BMR (basal metabolism rate). Recall that BMR refers to the energy expended in carrying out bodily functions like breathing, maintaining body temperature, etc. BMR rises immediately following intense and prolonged exercise and will remain elevated the following several hours. Over many weeks, however, daily vigorous exercise will build more lean muscle tissue, changing overall body composition. Since lean tissue is more metabolically active, there will be corresponding rise in BMR, resulting in better weight maintenance (Sjodin et al. 1996). Physically active individuals have good appetites, but exercising and eating are generally incompatible. To support exercise, the body must release glucose and fatty acids into the blood stream, as fuels, and simultaneously suppress its digestive functions. Recent research also suggests that aerobic exercises may cause levels of hormones like ghrelin to drop, thereby reducing appetite (Broom et al. 2009). Moreover, by displacing the act of eating, exercise may actually help curb appetite, especially if anxiety, stress, or depression are typical triggers of overeating in individuals (Epel et al. 2004).

13.19   Behavioral Change and Prevention Approaches to Overweight

The National Weight Control Registry, established in 1993, a collaborative venture between a number of universities, is a longitudinal prospective study of individuals 18 years and older, who have successfully maintained a 30 lb weight loss for a minimum of 1 year (Hill et al. 2005). Currently, the registry includes a data base of over 6000 individuals, who reported having made substantial changes in eating and exercise habits to lose weight and maintain their losses (Wing et al. 2012). For about 50% of the individuals who lost weight and maintaining the loss, without following any specific program, their success is attributed to behavioral change. This, of course, assumes that their self-reports of weight and weightloss are honest and accurate. Besides a fundamental emphasis on nutrition and exercise, behavioral programs which are designed to facilitate weight management typically include a number of strategies which rely on cognitive-behavioral change. These include "self-monitoring," "goal setting," "stimulus control," "problem-solving," "cognitive restructuring," and "relapse prevention" (Wing et al. 2012). For example, in self-monitoring program, the individuals are taught to write down everything they eat, the calorie content of the food they eat, as well as the grams of fat contained it. After a few weeks of weight loss program, the self-monitoring of physical activity is added. Advances in technology may soon replace paper and pencil methods, as devices such as personal data assistant (PDA) have been shown to improve adherence to self-monitoring in dietary interventions (Sevick et al. 2005). In contrast, stimulus control involves managing the near environment so as to avoid cues that encourage inappropriate eating or to institute new cues that elicit desirable behaviors. For the former, the individual may be instructed to place energy-dense foods out of sight (e.g., the office candy in a drawer rather than on one's desk) and, for the latter, to set up visible remainders to exercise (e.g., keeping sneakers near the doorway). Cognitive restructuring, in turn, refers to eliminating rationalizations for inappropriate eating, as well as countering negative thoughts with positive statements which build self-acceptance (Wing et al. 2012).

In conclusion, the importance of stress management in managing weight should be mentioned. Identifying cues which trigger overeating is a useful way to avoid temptation. Among binge eaters, in particular, chocolate is often eaten under the mistaken notion that it will help alleviate mental stress, as was briefly discussed earlier. However, it must be noted that instead of facilitating relaxation, and reducing depression, the net result of a chocolate binge may be an increase in feelings of guilt. Individuals who attempt to maintain weight loss and are fond of chocolate, then, may still partake of chocolate on occasion, but should avoid temptation of eating it, by keeping it around. Determining coping strategies, which do not involve food, such as engaging in exercise or watering houseplants, is best planned in anticipation of the stressful periods which inevitably occur from time to time. In all such cases, developing one's will power is, without doubt, the best option.

Given the difficulty of losing weight and maintaining the loss of weight, it should be obvious that preventing excessive weight gain in the first place is a sensible approach to lifelong health and well-being. Strategies for preventing weight gain are similar to approaches for losing weight, albeit they should begin early and be incorporated into daily routine. Members of the National Weight Control Registry have maintained, on average, a weight loss of 73 lb (33 kg) for almost 6 years. The key behavioral strategies that they report as helping them maintain their weight loss are listed in the following table (Table 13.5).

The dramatic increase in the prevalence of obesity, coupled with the persistent failure of most obesity treatment programs to achieve success, has resulted in the prevention research becoming one of the most important areas of enquiry in the field of obesity. If the goal of obesity prevention is defined as reducing the risk of it developing in adulthood, the treatment of obesity in children might be said to serve this purpose, albeit in a belated manner. Research suggests that programs utilizing the behavioral strategies previously addressed, which also involve parents in the treatment regimen, may be the most successful approach for working with the high-risk children. Indeed, experts agree that in treating children, treatment must be family centered and developmentally appropriate, and sound nutrition must be maintained to ensure normal linear growth – a phenomenon that makes setting weight goals for children all the more complicated.

As already mentioned, there has been increased discussion in recent years on the problem of living in an obesogenic, some would even say toxic, environment which

Table 13.5 Weight loss behaviors reported by members of the National Weight Control Registry

Consume a low-calorie, low-fat diet
Weigh self-daily
Maintain diet consistency across weekdays and weekends
Limit dietary variety
Eat breakfast daily
Limit fast-food intake
Use fat and sugar substitutes
Engage regularly in high levels of physical activity
Restrict television viewing

Source: Wing et al. (2012)

promotes overeating, poor nutrition habits, and sedentary behavior. In theory, there is a large range of public health strategies which could be applied to the problem of obesity. While education about health and nutrition will always be the mainstay of public health efforts, policy makers in countries like the USA are now considering taxing sugar-sweetened beverages or placing limitations on the advertising of certain energy foods, in particular, those commercials which are aimed at the children, such as high-sugar breakfast cereals. Far more funds are spent on promoting the consumption of foods of questionable nutritional value than on warnings to consumers about their hazards with respect to weight gain. The following example illustrates the point. While US $ 9.6 million sounds like a lot of money to promote the 5 A Day campaign to boost fruit and vegetable consumption, it pales in comparison to the US $ 11.3 billion spent on corporate advertising for soda, candy, snacks, and fast foods.

With special reference to children, designers of obesity prevention programs will do well to study the results in the following table (Table 13.6).

Latest Research on Obesity and Its Fallout   Latest research on obesity shows that the "nonalcoholic fatty liver disease" (NAFLD) in overweight elderly people leads to subsequent liver cirrhosis and may end up in cancer and malfunction of the liver.

The above guidelines will only apply to advanced countries such as the USA, Europe, and Scandinavia. Children's obesity is a real problem, in particular, in the USA, though it is spreading fast to other developed countries in the Western hemisphere. In the developing counties, such as India, it is still not as rampant as in the USA, but it is fast catching up among the affluent segment of the society, because of the economic upsurge, where children are pampered with fast foods or consuming them is becoming a fad among the young. Inasmuch as the underdeveloped countries in the continents, such as African, Latin American, or even some Asian countries like Pakistan or Sri Lanka, are concerned, it is hunger that is far more crucial. Hunger amidst plenty still rules, in pockets, but inviting health disorders by overeating is still a far cry there.

Table 13.6 Strategies to prevent child obesity

Adopt school curriculum that includes healthy eating, physical activity, and body image
Increase sessions for physical activity and the development of movement skills throughout the school week
Improve the nutritional quality of the food supply in schools
Ensure environments and cultural practices which support children eating healthier food and being physically active
Provide support for teachers and staff to implement health promotion activities
Support parents and home activities which encourage children to eat more nutritious foods, be more physically active, and engage in less television screen time

Source: Waters et al.

Aloe Vera: Blood Sugar Control, Uses, Side Effects, and Dosage

Got a recent diagnosis of diabetes from your doctor? Want to know the best holistic ways to treat this issue? Research shows approximately 30 million people are diabetes affected in India. Although a mounting problem in India as well as the world, it is quite easy to maintain and manage blood glucose levels with few herbal remedies and necessary routine changes

Modern medications today have proved insufficient for in-depth diabetes treatment. Therefore people have started opting for holistic herbal treatment that has been in existence for *thousands of years*. Aloe vera as an herbal supplement has minimal or no side effects for diabetes patients.

What Is Aloe Vera?

Aloe vera is a medicinal plant found extensively in the Arabian Peninsula. The *aloe vera plant* contains a clear gel that is packed with nutrients. It is used extensively in the cosmetic, pharmaceutical, and food industries.

Active Compounds in Aloe Vera for Diabetics

The prime component of aloe vera includes an outer green coverage with thorns like developments and an inner colorless gel. Leaves are the only part of aloe vera which are used for medicinal purposes. Aloe vera-based items are created from either of the components.

The plant houses more than 75 sets of active compounds which include enzymes, vitamins, minerals, monosaccharide, anthraquinones, lignin, polysaccharides, saponins, phytosterols, salicylic, as well as amino acids.

Additionally, the *aloe vera gel* also houses trace elements like magnesium, chromium, zinc, and manganese. These elements are crucial for the metabolism of glucose with improved insulin sensitivity. Aloe vera has also been known to address issues such as glaucoma, asthma, IBS, high blood pressure, and skin issues.

How Does Aloe Vera Affect Diabetes?

Research show aloe vera can aid in controlling *blood glucose levels*. Rich in antioxidants and vitamins, aloe is very effective in managing blood sugar levels. This is why aloe vera finds a special place in diabetes treatment.

Aloe vera has been associated with:

Increased insulin sensitivity:

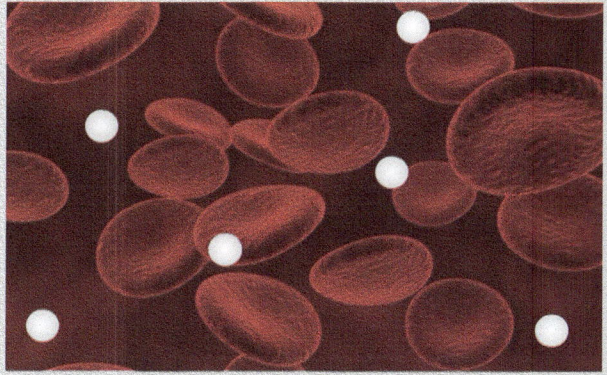

Packed with numerous bioactive components and antioxidants, aloe vera stimulates insulin secretion. It is therefore very helpful and in maintaining blood sugar levels.

Anti-inflammatory

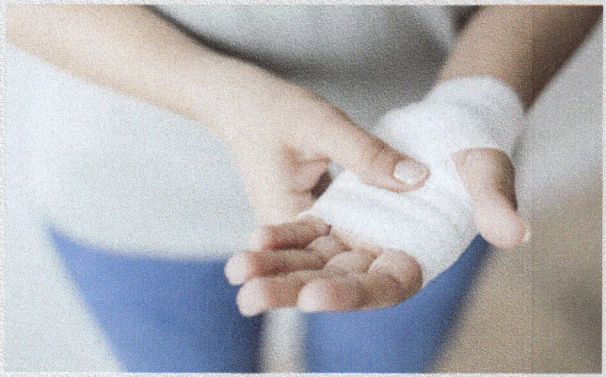

An incredible antioxidant, aloe vera juice houses anti-inflammatory properties protecting you from wounds, ulcers, and infections. People suffering from diabetes are vulnerable to infection and wounds that do not heal soon.

Hypoglycemic

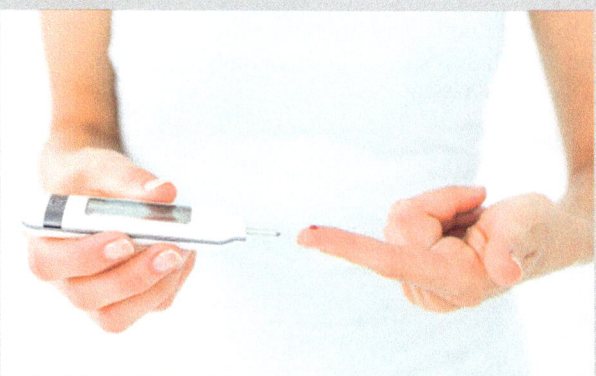

This particular property of aloe vera increases the ability of the tissues to respond to insulin and therefore can be beneficial for both NIDDM (non-insulin-dependent diabetes mellitus) and IDDM (insulin-dependent diabetes mellitus).

Less Fat

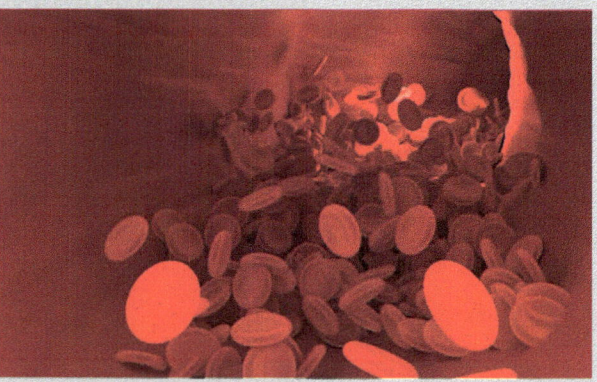

Helps to decrease levels of fats or blood lipids among patients suffering from abnormally high *blood sugar levels*.

Bioactive Compounds

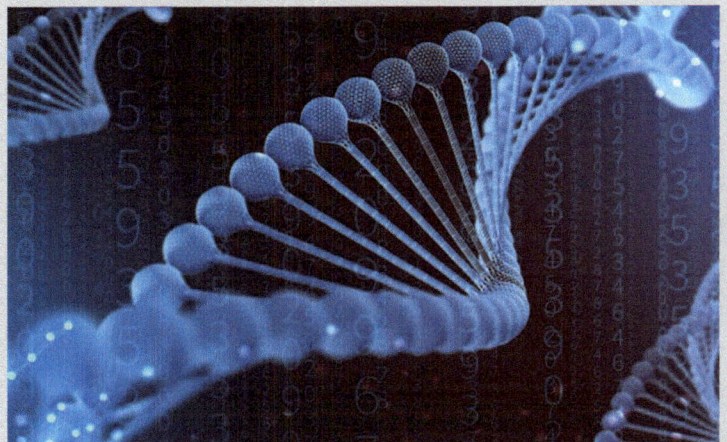

Aloe vera houses compounds like mannans, lectins, as well as anthraquinones that bind with carbohydrates in your body. This leads to a reduction in blood glucose levels.

Glucomannan

This dietary fiber is found in abundant quantity in *aloe vera gel*. It easily dissolves in water and reduces *blood sugar levels*.

Body Detoxification

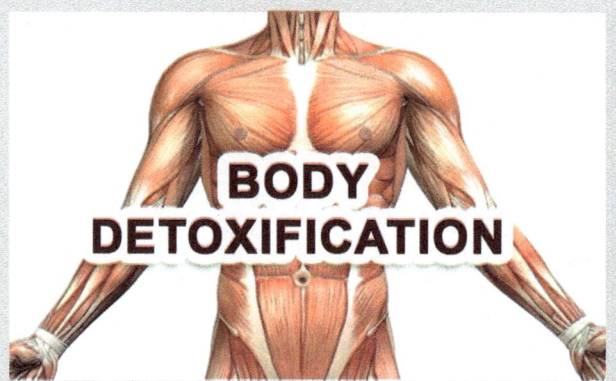

Additional *effects of aloe vera* include body detoxification which eliminates excess blood glucose.

Fast Healing

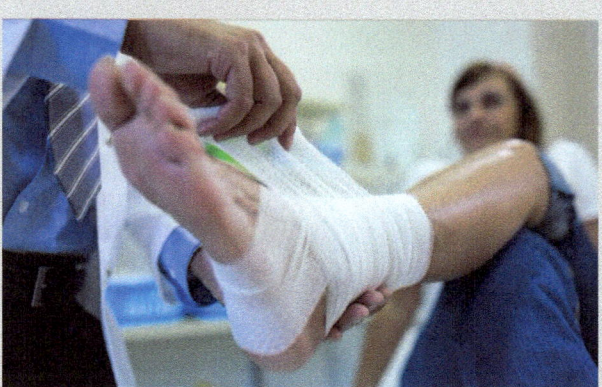

Decreased amount of swelling and faster process for healing wounds. Ulcers and leg wounds can be among the common diabetes complications that tend to take longer than normal to heal.

How to Use Aloe Vera in Our Daily Life?

Aloe vera in its formal form a colorless gel-like substance. However, it can be included in our daily routine in numerous forms:

- *Aloe vera juice* can be amazing for diabetes treatment when paired with a well-balanced healthy diet. Opt for fresh and organic *aloe vera juice* which is bitter in taste. So, if you cannot drink it raw, try and add a hint of honey to cut down the bitterness.
- 1 tablespoon fresh aloe vera juice twice a day regularizes blood insulin flow and releases constricted blood vessels for easy flow and protects the heart.
- Aloe vera gel and cream made out of its extracts are used as moisturizers and body lotion to treat acne, stretch marks, and sunburn.
- *Raw crushed leaves* of aloe vera is a form of treatment for diabetes.
- *Aloe vera gel* applied to diabetic boils reduces inflammation and irritation of the skin.
- Aloe vera gel treats minor cuts and scrapes to soothe them and promote healing.
- A bitter form of liquid termed as the aloe latex can be acquired from the *aloe leaf skin* which is used for things such as tablets, juice drinks, mouthwash, dental care, and capsules.
- **Side Effects of Aloe Vera**

Long-term unmonitored use of any product can have some drastic effects on our body. Especially consuming them in unnecessary high dosages can lead to some fatal outcomes:

Some common side effects of aloe include:

- Rare cases of hypoglycemia.

- In case you are allergic to any of its biocomponents, it can cause burning and itching of the skin.
- Regular intake of high dosage can cause stomach pain and cramps
- Long-term use at high doses will lead to diarrhea, kidney problems, blood in the urine, low potassium, muscle weakness, weight loss, and heart disturbances.
- **Bottom Line**

With years to come, it is evident that more than 50% of diabetics will move toward holistic herbal health treatment. So, whether you are looking for ways to *improve blood glucose levels* or cut down the symptoms of diabetes, aloe vera can be a great herb. It is very rich with antioxidants, vitamins, minerals, and bioactive components that aid in lowering and maintaining the blood glucose levels.

References

Ahima, R. S., Saper, C. B., Filler, J. S., & Elmquist, J. K. (2000). Leptin regulation of neuroendocrine systems. *Frontiers of Neuroendocrinology, 21*, 263–307.

Aronne, L. J., Neelinson, D. S., & Lillo, J. L. (2009). Obesity as a disease state: A new paradigm for diagnosis and treatment. *Clinical Cornerstone, 9*, 9–25.

Blundell, J. E., & MacDiamid, J. I. (1997). Fat as a risk factor for overconsumption: Situation, satiety and patterns of eating. *Journal of the American Dietetic Association, 97*, S63–S69.

Bray, G. A. (2008). Fructose: Should we worry? *International Journal of Obesity, 32*, S127–S131.

Bray, G. A. (2014). Obesity has always been with us: A historical introduction. In G. A. Bray & C. Bouchard (Eds.), *Handbook of obesity: Clinical applications* (3rd ed., pp. 3–18). Boca Raton: CRC Press.

Broom, D. R., Batterham, R. L., King, J. A., & Stensel, D. J. (2009). Influence of resistance and aerobic exercise on hunger, circulating levels of acylated ghrelin, and peptide YY in healthy males. *American Journal of Physiology-Regulatory, Integrative and Comparative Physiology, 296*, R29–R35.

Brownell, K. D., Kersh, R., Ludwig, D. S., Post, R. C., & Puhl, R. M. (2010). Personal responsibility and obesity: A constructive approach to a controversial issue. *Health Affairs, 29*, 379–387.

Burger, S., & Berner, L. A. (2014). A functional neuroimaging review of obesity, appetite hormones and ingestive behavior. *Physiology & Behavior, 136*, 121–127.

Burton-Freeman, B. (2000). Dietary fiber and energy regulation. *Journal of Nutrition, 130*, 272S–275S.

Cappuccio, F. P., Taggart, F. M., Kandala, N. B., Currie, A., & Peile, E. (2008). Meta-analysis of short sleep duration and obesity in children, adolescents and adults. *Sleep, 31*, 619–626.

Curioni, C. C., & Lourenco, P. M. (2005). Long - term weight loss after diet and exercise: A systematic review. *International Journal of Obesity, 29*, 1168–1174.

Darby, W. J., Ghalioungui, P., & Gravetti, L. (1977). *Food: The gift of Osiris*. London: Academic.

Das, S. K., & Roberts, S. B. (2012). Energy metabolism in fasting, fed, exercise, and re-feeding states. In J. W. Erdman, I. A. McDonald, & S. H. Zeisel (Eds.), *Present knowledge in nutrition* (10th ed., pp. 58–68). Ames: International Life Sciences Institute.

Despres, J. P., & Lemieux, I. (2006). Abdominal obesity and metabolic syndrome. *Nature, 444*, 881–887.

DiGirolamo, M., Harp, J., & Stevens, J. (2000). Obesity: Definition and epidemiology. In D. H. Lockwood & T. G. Heffner (Eds.), *Obesity: Pathology and therapy* (pp. 3–28). Berlin: Springer.

Economic Research Service. (2014). *Food availability (per capita) data system: Summary findings*. United States Department of Agriculture. Available at: http://www.ers.usda.gov/data-products/food-availability-per-capita-data-system/summary-findings.aspx. Accessed 25 Mar 2015.

Elia, M. (2001). Obesity in the elderly. *Obesity Research, 9*, 244S–248S.

Epel, E., Jimenez, S., Brownell, K., Stroud, L., Stonedy, C., & Niaaura, R. (2004). Are stress eaters at risk for the metabolic syndrome? *Annals of the New York Academy of Sciences, 1032*, 208–210.

Fernandez-Fernandez, R., & Tena-Sempere, M. (2013). Overview of ghrelin, appetite, and energy balance. In D. Bagchi & H. C. Preuss (Eds.), *Obesity: Epidemiology, pathophysiology and prevention* (2nd ed., pp. 161–169). Boca Raton: CRC Press.

Flegal, K. M., Shepherd, J. A., Looker, A. C., Graubard, B. I., & Borrud, L. G. (2009). Comparisons of percentage body fat, body mass index, waist circumference, and waist-stature ratio in adults. *American Journal of Clinical Nutrition, 89*, 500–508.

Fleury, C., Neverova, M., Collins, S., Raimbault, S., & Champigny, O. (1997). Uncoupling protein- 2: A novel gene linked to obesity and hyperinsulinemia. *Nature Genetics, 15*(3), 269–272.

Garcia, A. L., Wagner, K., Hothorn, T., Koebnick, C., Zunft, H. J. F., & Trippo, U. (2005). Improved prediction of body fat by measuring skinfold thickness, circumferences, and bone breadths. *Obesity, 13*, 626–634.

Goodpaster, B. H. (2002). Measuring body fat distribution and content in humans. *Current Opinion in Clinical Nutrition & Metabolic Care, 5*, 481–487.

Helmchen, L. A., & Henderson, R. M. (2004). Changes in the distribution of body mass index of white US men, 1890–2000. *Annals of Human Biology, 31*, 174–181.

Hill, J. O., Wyatt, H. R., Phelan, S., & Wing, R. R. (2005). The national weight control registry: Is it useful in helping deal with our obesity epidemic? *Journal of Nutrition Education and Behavior, 37*, 206–210.

Hu, F. B., Li, T. Y., Colditz, G. A., Willsett, W. C., & Manson, J. E. (2003). Television watching and other sedentary behaviors in relation to risk of obesity and type-2 diabetes mellitus in women. *Journal of the American Medical Association, 289*, 1785–1791.

Insel, P., Ross, D., McMahon, K., & Bernstein, M. (2013). *Discovering nutrition* (4th ed.). Burlington: Jones and Bartlett Learning.

Jain, A. (2004). *What works for obesity? A summary of the research behind obesity interventions*. London: BMJ Publishing Group.

Johansson, G., Wikman, A., Ahren, A. M., Hallmans, G., & Johansson, I. (2001). Underreporting of energy intake in repeated 24-hour recalls related to gender, age, weight status, day of interview, educational level, reported food intake, smoking habits and area of living. *Public Health Nutrition, 4*, 919–927.

Johnson, R. K., Appel, L. J., Brands, M., Howard, B. V., & Lefevre, M. (2009). Dietary sugars intake and cardiovascular health: A scientific statement from the American Heart Association. *Circulation, 120*, 1011–1020.

Kavey, R. E. (2010). How sweet it is: Sugar-sweetened beverage consumption, obesity, and cardiovascular risk in childhood. *Journal of the American Dietetic Association, 110*, 1450–1460.

Kirk, S. F. L., Penney, T. L., & McHugh, T. L. F. (2010). Characterizing the obesogenic environment: The state of the evidence with directions for future research. *Obesity Reviews, 11*, 109–117.

Knutson, K. L., Spiegel, K., Penev, P., & van Cauter, E. (2007). The metabolic consequences of sleep deprivation. *Sleep Medicine Reviews, 11*, 163–178.

Levine, J. A. (2004). Non-exercise activity thermogenesis (NEAT). *Nutrition Reviews, 62*, S82–S97.

Levine, J. A., Eberhardt, N. L., & Jensen, M. D. (1999). Role of non exercise activity thermogenesis in resistance in fat gain in humans. *Science, 283*, 212–214.

Levine, J. A., Vander Weg, M. W., Hill, J. O., & Klesges, R. C. (2006). Non-exercise activity thermogensesis: The crouching tiger hidden dragon societal weight gain. *Arteriosclerosis, Thrombosis, and Vascular Biology, 26*, 729–736.

Lucassen, E. A., Rother, K. I., & Gizza, G. (2012). Interacting epidemics? Sleep curtailment, insulin resistance, and obesity. *Annals of the New York Academy of Sciences, 1264*, 110–134.

Manco, M. (2013). Bariatric surgery and reversal of metabolic disorders. In D. Bagchi & H. C. Preuss (Eds.), *Obesity: Epidemiology, pathophysiology, and prevention* (2nd ed., pp. 931–945). Boca Raton: CRC Press.

Marketdata. (2014). *The US weight loss market: 2014 status report & forecast.* Marketdata Enterprises Inc. Available at: http://www.marketresearch.com/Marketdata-Enterprises-Inc-v416/Weight-Loss-Status-Forecast-8016030. Accessed 25 Mar 2015.

McElroy, S. L., Kotwal, R., Malhotta, S., Nelson, E. B., Keck, P. E., Jr., & Nemeroff, C. B. (2004). Are mood disorders and obesity related? A review for the mental health professional. *Journal of Clinical Psychiatry, 65*, 634–651.

McLaren, L. (2007). Socioeconomic status and obesity. *Epidemiologic Reviews, 29*, 29–48.

Miller, M. A., & Cappuccio, F. P. (2013). Sleep and obesity. In D. Bagchi & H. C. Preuss (Eds.), *Obesity: Epidemiology, pathophysiology, prevention* (2nd ed., pp. 179–190). Boca Raton: CRC Press.

Miyashita, K., & Hosokawa, M. (2013). Antiobesity by marine lipids. In D. Bagchi & H. C. Preuss (Eds.), *Obesity: Epidemiology, pathophysiology, and prevention* (2nd ed., pp. 651–665). Boca Raton: CRC Press.

Morton, N. M., Paterson, J. M., Masuzaki, H., Holmes, M. C., & Staels, B. (2004). Novel adipose tissue-mediated resistance to diet-induced visceral obesity in 11 β–hydroxysteroid dehydrogenase Type-1 deficient mice. *Diabetes, 53*, 931–938.

Mourao, D. M., Bressan, J., Campbell, W. W., & Mattes, R. D. (2007). Effects of food form on appetite and energy intake in lean and obese young adults. *International Journal of Obesity, 31*, 1688–1695.

Nestle, M., & Nesheim, M. (2012). *Why calories count: From science to politics.* Berkley: University of California Press.

Ng, M., Fleming, T., Robinson, M., Thomson, B., & Graetz, N. (2014). Global, national and regional prevalence of overweight and obesity in children and adults during 1980–2013: A systematic analysis for the Global Burden of Disease Study 2013. *The Lancet, 384*, 766–781.

Nieman, D. C. (2010). *Exercise testing and prescription* (7th ed.). New York: McGraw - Hill Higher Education.

Noakes, M., Keogh, J. B., Foster, P. R., & Clifton, P. M. (2005). Effect of an energy-restricted, high-protein, low-fat diet relative to a conventional high-carbohydrate, low-fat diet on weight loss, body composition, nutritional status, and markers of cardiovascular health in obese women. *American Journal of Nutrition, 81*, 1298–1306.

Paddon-Jones, D., Westman, E., Mattes, R. D., Wolfe, R. R., Astrup, A., & Weterterp-Plantenga, M. (2008). Protein, weight management and satiety. *American Journal of Clinical Nutrition, 87*, 1558S–1561S.

Pedersen, S. D., Sjodin, A., & Astrup, A. (2012). Obesity as a health risk. In J. W. Erdman, I. A. McDonald, & S. H. Zeisel (Eds.), *Present knowledge in nutrition* (10th ed., pp. 709–720). Ames: International Life Sciences Institute.

Pi-Sunyer, F. X. (2007). Weight management in diabetes prevention. In R. F. Kushner & D. H. Bessesen (Eds.), *Treatment of the obese patient* (pp. 243–263). Totowa: Humana Press Inc..

Popkin, B. M., Adair, L. S., & Ng, S. W. (2012). Global nutrition transition and the pandemic of obesity in developing countries. *Nutrition reviews, 70*, 3–21.

Puder, J. J., & Munsch, S. (2010). Psychological correlates of childhood obesity. *International Journal of Obesity, 34*, S37–S43.

Quetelet, L. A. J. (1871). *Anthropometrie ou mesure des differentes faculties de l'homme* [Anthropometry or measurement of different characteristics of man]. Brusels: Muqerdt

Rand, C. S., & MacGregor, A. M. (1991). Successful weight loss following obesity surgery and the perceived liability of morbid obesity. *International Journal of Obesity, 15*, 577–579.

Roberts, D. L., Dive, C., & Renehan, A. G. (2010). Biological mechanisms linking obesity and cancer risk: New perspectives. *Annual Review of Medicine, 61*, 310–316.

Rolls, B. J. (2003). The supersizing of America: Portion size and the obesity epidemic. *Nutrition Today, 38*, 42–53.

Rolls, B. J., Drewnowski, A., & Ledikwe, J. H. (2005). Changing the energy density of the diet as a strategy for weight management. *Journal of the American Dietetic Association, 105*, S98–S103.

Saris, W. H. M. (2001). Very-low –calorie diets and sustained weight loss. *Obesity, 9*, 295S–301S.

Schwartz, S. M., & Savastano, D. M. (2013). History and regulation of prescription and over-the-counter weight loss drugs. In D. Bagchi & H. C. Preuss (Eds.), *Obesity: Epidemiology, pathophysiology, and prevention* (2nd ed., pp. 313–348). Boca Raton: CRC Press.

Schwartz, M., Chambliss, H. O., Brownell, K., Blair, S., & Billington, C. (2003). Weight bias among health professionals specializing in obesity. *Obesity Research, 11*(9), 1033–1039.

Sevick, M. A., Piraino, B., Sereika, S., Starrett, T., & Bender, C. (2005). A preliminary study of PDA-based dietary self monitoring in hemodialysis patients. *Journal of Renal Nutrition, 15*, 304–311.

Sjodin, A. M., Forslund, A. N., Westerterp, H. N., Andersson, A. B., Forslund, J. M., & Hambraaeus, L. M. (1996). The influence of physical activity on BMR. *Medicine and Science in Sports and Exercise, 28*, 85–91.

Sturm, R. (2007). Increases in morbid obesity in the USA: 2000–2005. *Public Health, 121*, 492–496.

Sutin, A. R., & Terracciano, A. (2013). Perceived weight discrimination and obesity. *PLoS ONE, 87*, e70048. https://doi.org/10.1371/journal.pone.0070048.

Tuthill, A., Slawik, H., O'Rahilly, S., & Finer, N. (2006). Psychiatric co-morbidities in patients attending specialist obesity services in the UK. *Quarterly Journal of Medicine, 99*, 317–325.

Wang, Y., & Beydoun, M. A. (2007). The obesity epidemic in the United States – Gender, age, socioeconomic racial/ethnic, and geographic characteristics: A systematic review and meta-regression analysis. *Epidemiologic Reviews, 29*, 6–28.

Wing, R. R., Gorin, A., & Tate, D. F. (2012). Strategies for changing eating and exercise behavior to promote weight loss and maintenance. In J. W. Erdman, I. A. McDonald, & S. H. Zeisel (Eds.), *Present knowledge in nutrition* (10th ed., pp. 1057–1070). Ames: International Life Sciences Institute.

Worobey, J. (2002). Interpersonal versus intrafamilial predictors of maladaptive eating attitudes in young women. *Social Behavior and Personality, 30*, 424–434.

Chapter 14
Conclusion

Abstract It is an overview of the various contents of the book.

Keywords "Live to eat" · "Eat to live" · Food · Human behavior

This book which contains 14 chapters discusses the various aspects of food and human behavior. Many a time, we underestimate the crucial role food plays on one's behavior/character. One observes many "who live to eat," and some others, "who eat to live." There is a world of difference between the two, as the readers would note from details given in Chap. 13, which chronicles the outcome of the first paradox of life "living to eat." To keep the record straight, one would observe that many of modern-day illnesses could be traced to the wrong eating habit one develops over the years, in one's lifetime. Indian sages are known, as *Vedic* books would reveal, to have lived without food for days on end, just subsisting on water. The most illustrative example is set by India's Prime Minister, Shri Narendra Modi, who observes a fast during 9 days in the holy month of October during *Durga Puja*, where Goddess Durga, who symbolizes the triumph of truth over evil, is worshipped. Interestingly, during these 9 days, each year, since many past, Shri Modi undertakes this fast. Even during his physically very strenuous and demanding trip to the USA, immediately after he was sworn in as Prime Minister of India, he maintained his 9 days *vrath* (the word *vrath* in Sanskrit means fast), subsisting on just fresh lemon juice and plain hot water. The body reacts to the fast. The latest medical evidence shows that even among diabetics, continuous fast over several hours triggers a reaction in the pancreas, which overcomes the "insulin resistance" – the cause being the pancreatic cells become more vigilant and active to adapt to the fasting. Such examples could be multiplied.

I have drawn upon the published work of several of my colleagues working in the area of human nutrition-related behavioral responses. I owe all of them a deep debt of gratitude and to you reader, for that leap of faith, in picking up this book, as the validation of the firm belief of a dedicated and untiring scientist, in the dictum "we are what we eat, let us be healthy and happy in choosing the right food."

© Springer Nature Switzerland AG 2020
K. P. Nair, *Food and Human Responses*,
https://doi.org/10.1007/978-3-030-35437-4_14

References

Abbot, N. J., Ronnback, L., & Hansson, E. (2006). Astrocyte-endothelial interactions at the blood-brain barrier. *Nature Reviews Neuroscience, 7*, 41–53.

Abbott, M., & Goodheart, K. L. (2012). Cognitive behavioral approaches for treating eating disorders. In K. L. Goodheart, J. R. Clopton, & J. J. Robert-McComb (Eds.), *Eating disorders in women and children* (pp. 371–383). Boca Raton: CRC Press.

Abel, E. L. (1984). *Fetal alcohol syndrome and fetal alcohol effects.* New York: Plenum Press.

Ahima, R. S., Saper, C. B., Filler, J. S., & Elmquist, J. K. (2000). Leptin regulation of neuroendocrine systems. *Frontiers of Neuroendocrinology, 21*, 263–307.

Alaimo, K., Olson, C., & Frongillo, E. A. (2002). Food insufficiency, but not low family income, is positively associated with dysthymia and suicide symptoms in adolescents. *Journal of Nutrition, 132*, 719–725.

Alberts, B., Johnson, A., Lewis, J., Raff, M., Roberts, K., & Walter, P. (2008). *Molecular biology of the cell* (5th ed.). New York: Garland Science.

Albu, J., & Pi-Sunyer, F. X. (2003). Obesity and diabetes. In G. A. Bray & C. Bouchard (Eds.), *Handbook of obesity: Etiology and pathophysiology* (2nd ed., pp. 697–708). New York: Marcel Dekker.

Amer, A., Breu, J., McDermott, J., Wurtman, R. J., & Maher, T. J. (2004). 5-hydroxy-L-tryptophan suppresses food intake in food-deprived and stressed rats. *Pharmacology, Biochemistry & Behavior, 77*(1), 137–143.

Anderson, G. H., & Hrboticky, N. (1986). Approaches to assessing the dietary component of diet-behavior connection. Nutrition reviews. *Diet and Behavior: A Multidisciplinary Evaluation, 44*, 42–51.

Anderson, J. W., Johnstone, B. M., & Remley, D. T. (1999). Breast-feeding and cognitive development: A meta-analysis. *American Journal of Clinical Nutrition, 70*, 525–535.

Andrews, H. (2012). The assessment of mental state, psychiatric risk, and co-morbidity in eating disorders. In J. R. E. Fox & K. P. Goss (Eds.), *Eating and its disorders* (pp. 11–27). Malden: Wiley.

Aquilonius, S., & Eckeras, S. (1977). Choline therapy in Huntington's chorea. *Neurology, 27*, 887–889.

Arnaud, M. J. (2011). Pharmacokinetics and metabolism of natural methylxanthines in animal and man. *Handbook of Experimental Pharmacology, 200*, 33–91.

Aronne, L. J., Neelinson, D. S., & Lillo, J. L. (2009). Obesity as a disease state: A new paradigm for diagnosis and treatment. *Clinical Cornerstone, 9*, 9–25.

Avena, N. M., Rada, P., & Hoebel, B. G. (2008). Evidence for sugar addiction: Behavioral and neurochemical effects of intermittent, excessive sugar intake. *Neuroscience and Biobehavioral Reviews, 32*, 20–39.

© Springer Nature Switzerland AG 2020

K. P. Nair, *Food and Human Responses*,

https://doi.org/10.1007/978-3-030-35437-4

Azevedo, F. A. C., Carvalho, L. R. B., Grinberg, L. T., Farfel, J. M., Ferretti, R. E. L., Leite, R. E. P., Filho, W. J., Lent, R., & Herculano-Houzel, S. (2009). Equal numbers of neuronal and nonneuronal cells make the human brain as isometrically scaled-up primate brain. *Journal of Comparative Neurology, 513*, 541.

Baker, R. D., Greer, F. R., & the Committee on Nutrtion. (2010). Clinical report – Diagnosis and prevention of iron deficiency and iron-deficiency anemia in infants and young children (0–3 years of age). *Pediatrics, 126*, 1040–1050.

Banderet, L. E., & Lieberman, H. R. (1989). Treatment with tyrosine, a neurotransmitter precursor; reduces environmental stress in humans. *Brain Research Bulletin, 22*, 759–762.

Barker, D. J. P. (Ed.). (1992). *Fetal and infant origins of adult disease*. London: BMJ.

Barrett, B., Kiefer, D., & Rabago, D. (1999). Assessing the risks and benefits of herbal medicine: An overview of scientific evidence. *Alternative Therapies in Health and Medicine, 5*, 40–49.

Bartel, P. R., Griessel, R. D., Burnett, L. S., Freiman, I., Rosen, E. U., & Geefhuysen, J. (1978). Long-term effects of kwashiorkor on psychomotor development. *South African Medical Journal, 53*, 360–362.

Bautista, A., Barker, P. A., Dunn, J. T., Sanchez, M., & Kaiser, D. L. (1982). The effects of oral iodized oil on intelligence, thyroid status, and somatic growth in school-age children from an area of endemic goiter. *American Journal of Clinical Nutrition, 35*, 127–134.

Bayley, N. (1993). *Bayley scales of infant development* (2nd ed.). San Antonio: Psychological Corporation.

Bazan, N. G. (1990). Supply of n-3 polyunsaturated fatty acids and their significance in the central nervous system. In R. J. Wurtman & J. J. Wurtman (Eds.), *Nutrition and the brain* (Vol. 8, pp. 1–24). New York: Raven Press.

Bell, D. G., & McLellan, T. M. (2003). Effect of repeated caffeine ingestion on repeated exhaustive exercise endurance. *Medicine & Science. Sports & Exercise, 35*, 1348–1354.

Bellisle, F. (2004). Effects of diet on behavior and cognition in children. *British Journal of Nutrition, 92*, S227–S232.

Bent, S., Goldberg, H., Padula, A., & Avins, A. L. (2005). Spontaneous bleeding associated with Ginkgo biloba: A case report and systematic review of the literature. *Journal of General Internal Medicine, 20*, 657–661.

Bent, S., Padula, A., Moore, D., Patterson, M., & Mehling, W. (2006). Valerian for sleep: A systematic review and meta-analysis. *American Journal of Medicine, 119*, 1005–1012.

Benten, D. (1997). Dietary fat and cognitive functioning. In M. Hillbrand & T. Spitz (Eds.), *Lipids, health, and behavior* (pp. 227–243). Washington, DC: American Psychological Association.

Benton, D. (2002). Selenium intake, mood and other aspects of psychological functioning. *Nutritional Neuroscience, 5*, 363–374.

Benton, D. (2003). Carbohydrate, memory and mood. *Nutrition Reviews, 61*, S61–S67.

Bernstein, G. A., Carroll, M. E., Thuras, P. D., Cosgrove, K. P., & Roth, M. E. (2002). Caffeine dependence in teenagers. *Drug and Alcohol Dependence, 66*, 1–6.

Birch, E. E., Garfield, S., Hoffman, D. R., Uauy, R., & Birch, D. G. (2000). A randomized control trial of early dietary supply of long chain polyunsaturated fatty acids and mental development in term infants. *Developmental Medicine and Child Neurology, 42*, 174–181.

Blanchard, J., Weber, C. W., & Shearer, L. E. (1992). Methylxanthine levels in breast milk of lactating women of different ethnic and socioeconomic classes. *Biopharmaceutics & Drug Disposition, 13*, 187–196.

Blass, E. M., & Shah, A. (1995). Pain-reducing properties of sucrose in human newborns. *Chemical Senses, 20*, 29–35.

Blass, E. M., & Smith, B. A. (1992). Differential effects of sucrose, fructose, glucose, and lactose on crying in 1–3 day-old human infants: Qualitative and quantitative considerations. *Developmental Psychology, 28*, 804–810.

Blass, E. M., Fitzgerald, E., & Kehoe, P. (1987). Interactions between sucrose, pain, and isolation distress. *Pharmacology, Biochemistry & Behavior, 26*, 483–489.

Bleichrodt, N., & Born, M. (1994). A meta-analysis of research on iodine and its relationship to cognitive development. In J. Stanbury (Ed.), *The damaged brain of iodine deficiency.* New York: Cognizant Communication.

Bloom, R. E., Nelson, C. A., & Lazerson, A. (2006). *Brain, mind and behavior* (3rd ed.). New York: W.H. Freeman.

Blundell, J. E., & Hill, A. J. (1987). Influence of tryptophan on appetite and food selection in man. In S. Kaufmann (Ed.), *Amino acids in health and disease: New perspectives* (pp. 403–419). New York: Alan R, Liss.

Blundell, J. E., & MacDiamid, J. I. (1997). Fat as a risk factor for overconsumption: Situation, satiety and patterns of eating. *Journal of the American Dietetic Association, 97*, S63–S69.

Boden, J. M., & Fergusson, D. M. (2011). Alcohol and depression. *Addiction, 106*, 906–914.

Booth, D. A. (1994). *Psychology of nutrition.* London: Taylor & Francis.

Boulenger, J. P., Patel, J., Post, R. M., Parema, A. M., & Marangos, P. J. (1983). Chronic caffeine consumption increases the number of brain adenosine receptors. *Life Sciences, 32*, 1135–1142.

Boyle, T. C. (1984). *The road to Wellville.* London: Penguin Books.

Bradley, R. H., & Corwyn, R. F. (2002). Socioeconomic status and child development. *Annual Review of Psychology, 53*, 371–399.

Bray, G. A. (2008). Fructose: Should we worry? *International Journal of Obesity, 32*, S127–S131.

Bray, G. A. (2014). Obesity has always been with us: A historical introduction. In G. A. Bray & C. Bouchard (Eds.), *Handbook of obesity: Clinical applications* (3rd ed., pp. 3–18). Boca Raton: CRC Press.

Bray, G. A., Lovejoy, J. C., Most-Windhauser, M., Smith, S. R., Voaufiva, J., Denkins, Y., deJonge, L., Rood, J., Lefvre, M., Eldridge, A. L., & Peters, J. C. (2002). A 9-mo randomized clinical trial comparing fat-substituted and fat-reduced diets in healthy obese men: The Ole Study. *American Journal of Clinical Nutrition., 76*, 928–934.

Broft, A., Berner, L. A., & Walsh, T. B. (2010). Pharmacotherapy for bulimia nervosa. In C. M. Grilo & J. E. Mitchell (Eds.), *The treatment of eating disorders: A clinical handbook* (pp. 388–401). New York: Guilford Press.

Broom, D. R., Batterham, R. L., King, J. A., & Stensel, D. J. (2009). Influence of resistance and aerobic exercise on hunger, circulating levels of acylated ghrelin, and peptide YY in healthy males. *American Journal of Physiology-Regulatory, Integrative and Comparative Physiology, 296*, R29–R35.

Brown, J. L., & Pollitt, E. (1996). Malnutrition, poverty, and intellectual development. *Scientific American, 274*, 38–43.

Brownell, K. D., Kersh, R., Ludwig, D. S., Post, R. C., & Puhl, R. M. (2010). Personal responsibility and obesity: A constructive approach to a controversial issue. *Health Affairs, 29*, 379–387.

Bruch, H. (1973). *Eating disorders: Obesity, anorexia nervosa, and the person within.* New York: Basic Books.

Brunye, T. T., Mahoney, C. R., Lieberman, H. R., & Taylor, H. A. (2010). Caffeine modulates attention network function. *Brain and Cognition, 72*, 181–188.

Bulik, C. M., Berkman, N. D., Brownley, K. A., Sedway, J. A., & Lohr, K. N. (2007). Anorexia nervosa treatment: A systematic review of randomized controlled trials. *International Journal of Eating Disorders, 40*, 310–320.

Bunn, H. T., & Ezzo, J. A. (1993). Hunting and scavenging by Plio-Pleistocene hominids: Nutritional constraints, archaeological patterns, and behavioural implications. *Journal of Archaelogical Science, 20*, 365–398.

Burger, S., & Berner, L. A. (2014). A functional neuroimaging review of obesity, appetite hormones and ingestive behavior. *Physiology & Behavior, 136*, 121–127.

Burton-Freeman, B. (2000). Dietary fiber and energy regulation. *Journal of Nutrition, 130*, 272S–275S.

Cantrell, P. A. (2000). Beer and ale. In K. F. Kiple & K. C. Ornelas (Eds.), *The Cambridge world history of food* (pp. 619–625). Cambridge: Cambridge University Press.

Cappuccio, F. P., Taggart, F. M., Kandala, N. B., Currie, A., & Peile, E. (2008). Meta-analysis of short sleep duration and obesity in children, adolescents and adults. *Sleep, 31*, 619–626.

Carlsen, M. H., Halvorsen, B. L., Holte, K., Bohn, S. K., & Draglund, S. (2010). The total antioxidant content of more than 3100 foods, beverages, spices, herbs and supplements used worldwide. *Nutrition Journal, 9*, 3.

Carlsson, C. M., Gleason, C. E., Hess, T. M., Moreland, K. A., & Blazel, H. L. (2008). Effects of simvastatin on cerebrospinal fluid biomarkers and cognition in middle -aged adults at risk for Alzheimer's disease. *Journal of Alzheimer's Disease, 13*, 187–192.

Castellanos, F. X., & Rapoport, J. L. (2002). Effects of caffeine on development and behavior in infancy and childhood: A review of the published literature. *Food and Chemical Toxicology, 40*, 1235–1242.

Catalan, J., Toru, M., Slotnick, B., Murthy, M., Grener, R. S., & Salem, N., Jr. (2002). Cognitive deficits in docosahexaenoic acid-deficient rats. *Behavioral Neuroscience, 116*, 1022–1031.

Caton, S. J., Ball, M., Aherm, A., & Hetherington, M. M. (2004). Dose-dependent effects of alcohol and appetite and food intake. *Physiology and Behavior, 81*, 51–58.

Centers for Disease Control and Prevention. (2013). Attention Deficit/Hyperactivity Disorder (ADHD). *Data & Statistics*. Available at: http://www.cdc.gov/ncbddd/adhd/data.html. Accessed 25 Mar 2015.

Chanhangeux, J. P., & Ricoeur, P. (2000). *What makes us think?* Princeton: Princeton University Press.

Charles, H. C., Lazeyras, F., Krishnan, K. R. R., Boyko, O. B., Payne, M., & Moore, D. (1994). Brain choline in depression: In vivo detection of potential pharmacodynamic effects of antidepressant therapy using hydrogen localized spectroscopy. *Progress in Neuropsychopharmacology & Biological Psychiatry, 118*, 1121–1127.

Chew, M. L., Mulsant, B. H., Pollock, B. G., Lehman, M. E., & Greenspan, A. (2008). Anticholinergic activity of 107 medications commonly used by older adults. *Journal of the American Geriatrics Society, 56*, 1333–1341.

Chial, H. J., McAlpine, D. E., & Camilleri, M. (2002). Anorexia nervosa: Manifestations and management for the gastroenterologist. *The American Journal of Gastroenterology, 97*, 255–269.

Christensen, L. (1996). *Diet-behavior relationships: Focus on depression*. Washington, DC: American Psychological Association.

Christensen, L., & Pettijohn, L. (2001). Mood and carbohydrate cravings. *Appetite, 36*, 137–145.

Christofides, A., Asante, K. P., Schauer, C., Sharief, W., Owusu-Agyei, S., & Zlotkin, S. (2006). Multi-micronutrient Sprinkles including a low dose of iron provided as microencapsulated ferrousfumarate improves haematologic indices in anaemic children: A randomized clinical trial. *Maternal and Child Nutrition, 2*, 169–180.

Cockfield, A., & Philpot, U. (2009). Feeding size 0: The challenges of anorexia nervosa. Managing anorexia from a dietitian's perspective. *Proceedings of the Nutrition Society, 68*, 281–288.

Coggins, T. E., Oiswant, L. B., Olson, H. C., & Timler, G. R. (2003). On becoming social competent communicators: The challenge for children with fetal alcohol exposure. *International Review of Research in Mental Retardation, 27*, 121–150.

Cohen, P. A. (2012). DMAA as a dietary supplement ingredient. *Archives of Internal Medicine, 172*, 1038–1039.

Colen, C. G., & Ramey, D. M. (2014). Is breast truly best? Estimating the effects of breastfeeding on long-term child health and wellbeing in the United States using sibling comparisons. *Social Science & Medicine, 109*, 55–65. https://doi.org/10.1016/j.socscimed.2014.01.027.

Connor, W. E., Neuringer, M., & Reisbick, S. (1992). Essential fatty acids. The importance of n-3 fatty acids in the retina and brain. *Nutrition Review, 50*, 21–29.

ConsumerLab. com. (2010). *Survey of vitamin and supplement users*. Available at: http://www.consumerlab.com/news/Supplement_Survey_Report/1_31_2010. Accessed 25 Mar 2015

Corell, C. U., Leucht, S., & Kane, J. (2004). Lower risk for tardive dyskinesia associated with second-generation anti-psycholics: A systematic review of 1-year studies. *American Journal of Psychiatry, 161*, 414–425.

Corsica, J. A., & Spring, B. J. (2008). Carbohydrate craving: A double-blind, placebo-controlled test of the self medication hypothesis. *Eating Behaviors, 9*, 447–454.

Cosman, M. P. (1983). A feast of Aesculapius: Historical diets for asthma and sexual pleasure. *Annual Review of Nutrition, 3*, 1–33.

Cravioto, J. (1977). Not by bread alone: Effect of early malnutrition and stimuli deprivation on mental development. In O. P. Ghai (Ed.), *Perspectives in pediatrics* (pp. 87–104). New Delhi: Interprint.

Cravioto, J., & Arrieta, R. (1986). Nutrition, mental development, and learning. In F. Faulkner & J. M. Tanner (Eds.), *Human growth* (Vol. 3, pp. 501–536). New York: Plenum Publishing.

Crow, S. J., Peterson, C. B., Swanson, S. A., Raymond, N. C., & Specker, S. (2009). Increased mortality in bulimia nervosa and other eating disorders. *American Journal of Psychiatry, 16*, 173–176.

Cunnane, S. C., Plourde, M., Pifferi, F., Begin, M., Feart, C., & Barberger-Gateau, P. (2009). Fish, docosahexaenoic acid and Alzheimer's disease. *Progress in Lipid Research, 48*, 239–256.

Curioni, C. C., & Lourenco, P. M. (2005). Long - term weight loss after diet and exercise: A systematic review. *International Journal of Obesity, 29*, 1168–1174.

Daly, J. W., & Fredholm, B. B. (1998). Caffeine- an atypical drug of dependence. *Drug And Alcohol dependence, 51*, 199–206.

Dangour, A. D., & Allen, E. (2013). Do omega-3 fat boost brain function in adults? Are we any closer to an answer? *American Journal of Clinical Nutrition, 97*, 909–910.

Darby, W. J., Ghalioungui, P., & Gravetti, L. (1977). *Food: The gift of Osiris*. London: Academic.

Das, S. K., & Roberts, S. B. (2012). Energy metabolism in fasting, fed, exercise, and re-feeding states. In J. W. Erdman, I. A. McDonald, & S. H. Zeisel (Eds.), *Present knowledge in nutrition* (10th ed., pp. 58–68). Ames: International Life Sciences Institute.

Davidson, M., Stern, R. G., Bierer, L. M., Horvath, T. B., Zemishlani, Z., Markofsky, R., & Mohs, C. S. (1991). Cholinergic strategies in the treatment of Alzheimer's disease. *Acta Psychiatrica Scandinavica, 366*, 47–51.

Davis, J. M., Zhao, Z. W., Stock, H. S., Mehl, K. A., Buggy, J., & Hand, G. A. (2003). Central nervous system effects of caffeine and adenosine on fatigue. *American Journal of Physiology: Regulatory, Integrative and Comparative Physiology, 284*, R399–R404.

Delange, F. (2000). The role of iodine in brain development. *Proceedings of the Nutrition Society, 59*, 75–79.

Delgado, P. L., Price, L. H., Miller, H. L., Salomon, R. M., Licinio, J., Krystal, J. H., Heninger, G. R., & Charney, D. S. (1991). Rapid serotonin depletion as a provocative challenge test for patients with major depression: Relevance to antidepressant action and the neurobiology of depression. *Psychopharmacology Bulletin, 27*, 321–329.

Despres, J. P., & Lemieux, I. (2006). Abdominal obesity and metabolic syndrome. *Nature, 444*, 881–887.

Devore, E. E., Grodstein, F., van Rooij, F. J. A., Hofman, A., & Rosner, B. (2009). Dietary intake of fish and omega-3 fatty acids in relation to long-term dementia risk. *American Journal of Clinical Nutrition, 90*, 170–176.

Diagnostic and Statistical Manual of Mental Disorders. 5th edn (DSM-5). (2013). American Psychiatric Association, Washington, DC.

Diagnostic and Statistical Manual of Mental Disorders (DSM-5). (2013). American Psychiatric Association, Arlington, Virginia.

Dietschy, J. M., & Turley, S. D. (2004). Cholesterol metabolism in the central nervous system during early development and in the mature animal. *Journal of Lipid Research, 45*, 1375–1397.

DiGirolamo, M., Harp, J., & Stevens, J. (2000). Obesity: Definition and epidemiology. In D. H. Lockwood & T. G. Heffner (Eds.), *Obesity: Pathology and therapy* (pp. 3–28). Berlin: Springer.

Dover, G. J. (2009). The Barker-Hypothesis: How pediatricians will diagnose and prevent common adult-onset of diseases. *Transactions of the American Clinical and Climatological Association, 120*, 199–207.

Economic Research Service. (2014). *Food availability (per capita) data system: Summary findings.* United States Department of Agriculture. Available at: http://www.ers.usda.gov/data-products/food-availability-per-capita-data-system/summary-findings.aspx. Accessed 25 Mar 2015.

Elia, M. (2001). Obesity in the elderly. *Obesity Research, 9,* 244S–248S.

Ellenbogen, M. A., Young, S. N., Dean, P., Palmour, R. M., & Benkelfat, C. (1996). Mood response to acute tryptophan depletion in healthy volunteerts: Sex differences and temporal stability. *Neuropharmacology, 15*(5), 465–474.

Endres, J. B., Rockwell, R. B., & Mense, C. G. (2004). *Food, nutrition, and the young child.* New York: Macmillan Publishing Company.

Engelberg, H. (1992). Low serum cholesterol and suicide. *The Lancet, 339,* 727–729.

Engelhart, M. J., Geelings, M. I., Ruitenberg, A., van Swieten, J. C., Hofman, A., Witteman, J. C., & Breteler, M. M. (2002). Dietary intake of antioxidants and risk of Alzheimer disease. *Journal of the American Medical Association, 287,* 3223–3229.

Enslen, M., Milon, H., & Malone, A. (1991). Effect of low intake of n-3 fatty acvids during development on brain phospholipids fatty acid composition and exploratory behavior in rats. *Lipids, 26,* 203–208.

Epel, E., Jimenez, S., Brownell, K., Stroud, L., Stonedy, C., & Niaaura, R. (2004). Are stress eaters at risk for the metabolic syndrome? *Annals of the New York Academy of Sciences, 1032,* 208–210.

Ereshefsky, L., Rospod, R., & Jann, M. (1989). Organic brain syndromes, Alzheimer type. In J. T. DiPiro, R. L. Talbert, P. E. Hayes, G. C. Yee, & L. M. Posey (Eds.), *Pharmacotherapy: A pathophysiological approach* (pp. 678–696). New York: Elsevier.

Ernst, E. (2002). The risk-benefit profile of commonly used herbal therapies: Ginkgo, St John's Wort, Ginseng, Echinacea, Saw Palmetto, and Kava. *Annals of Internal Medicine, 136,* 42–53.

Espinosa, M. P., Sigman, M. D., Neuman, C. G., Bwibo, N. O., & McDonald, M. A. (1992). Playground behaviors of school-age children in relation to nutrition, schooling, and family characteristics. *Developmental Psychology, 28,* 1188–1195.

Etminan, M., Gill, S. S., & Samii, A. (2005). Intake of vitamin E, vitamin C, and carotenoids and the risk of Parkinson's disease: A meta-analysis. *The Lancet Neurology, 4,* 362–365.

Fagan, J. F., & Shephers, P. A. (1987). *The Fagan test of infant intelligence.* Cleveland: Infantest Corporation.

Fahn, S. (1991). An open trial of high-dosage antioxidants in early Parkinson's disease. *American Journal of Clinical Nutrition, 53,* 380S–382S.

Farb, P., & Armelagos, G. (1980). *Consuming passions: The anthropology of eating.* Boston: Houghton-Mifflin.

Farris, M. W., & Zhang, J. G. (2003). Vitamin E therapy in Parkinson's disease. *Toxicology, 189,* 129–146.

Feng, L., Yap, K. B., Kua, E. H., & Ng, T. P. (2010). Statin use and depressive symptoms in a prospective study of community-living older persons. *Pharmacoepidemiology and Drug Safety, 19,* 942–948.

Fenten, W. S., Dickerson, F., Boronow, J., Hibbeln, J. R., & Knable, M. (2001). A placebo-controlled trial of omega 3-fatty acid (ethyl eicosapentaenoic acid) supplementation for residual symptoms and cognitive impairment in schizophrenia. *American Journal of Psychiatry, 158,* 2071–2074.

Fernandez-Fernandez, R., & Tena-Sempere, M. (2013). Overview of ghrelin, appetite, and energy balance. In D. Bagchi & H. C. Preuss (Eds.), *Obesity: Epidemiology, pathophysiology and prevention* (2nd ed., pp. 161–169). Boca Raton: CRC Press.

Fernstrom, J. D. (2000). Can nutrient supplements modify brain function? *American Journal of Nutrition, 7,* 1669S–1673S.

Fisone, G., Borgkvist, A., & Usiello, A. (2004). Caffeine as a psychomotor stimulant: Mechanism of action. *Cellular and Molecular Life Sciences, 61,* 857–872.

Flegal, K. M., Shepherd, J. A., Looker, A. C., Graubard, B. I., & Borrud, L. G. (2009). Comparisons of percentage body fat, body mass index, waist circumference, and waist-stature ratio in adults. *American Journal of Clinical Nutrition, 89,* 500–508.

Fleury, C., Neverova, M., Collins, S., Raimbault, S., & Champigny, O. (1997). Uncoupling protein- 2: A novel gene linked to obesity and hyperinsulinemia. *Nature Genetics, 15*(3), 269–272.

Food and Nutrition Board, Institute of Medicine and National Academy of Sciences. (2000). *Dietary reference intakes for vitamin C, vitamin E, selenium and carotinoids.* Washington, DC: National Academy Press.

Food and Nutrition Board, Institute of Medicine, National Academy of Sciences. (2001a). Iron. In *Dietary reference intakes for vitamin A, vitamin K, arsenic, boron, chromium, copper, iodine, iron, manganese, molybdenum, nickel, silicon, vanadium and zinc* (pp. 290–393). Washington, DC: National Academy Press.

Food and Nutrition Board, Institute of Medicine, National Academy of Sciences. (2001b). Zinc. In *Dietary reference intakes for vitamin A, vitamin K, arsenic, boron, chromium, copper, iodine, iron, manganese, molybdenum, nickel, silicon, vanadium, and zinc* (pp. 442–501). Washington, DC: National Academy Press.

Food and Nutrition Board, Institute of Medicine, National Academy of Sciences. (2001c). Iodine. In *Dietary reference intakes for vitamin A, vitamin K, arsenic, boron, chromium, copper, iodine, iron, manganese, molybdenum, nickel, silicon, vanadium, and zinc* (pp. 258–289). Washington, DC: National Academy Press.

Forgo, I., & Schimert, G. (1985). The duration of effect of the standardized ginseng extract G 115 in healthy competitive athletes. *Notabene Medici, 15,* 636–640.

Fugh-Berman, A., & Cott, J. M. (1999). Dietary supplements and natural products as psychotherapeutic agents. *Psychosomatic Medicine, 61,* 712–728.

Furham, A., & Alibhai, N. (1983). Cross-cultural differences in the perception of female body shapes. *Physiological medicine, 13,* 829–837.

Furst, T., Connors, M., Bisogni, C. A., Sobal, J., & Falk, L. W. (1996). Food choice: A conceptual model of the process. *Appetite, 26,* 247–265.

Galler, J. R., & Ramsey, F. (1985). The influence of early malnutrition on subsequent behavioral development: The role of the micro environment of the household. *Nutrition and behavior, 2,* 161–173.

Garcia, A. L., Wagner, K., Hothorn, T., Koebnick, C., Zunft, H. J. F., & Trippo, U. (2005). Improved prediction of body fat by measuring skinfold thickness, circumferences, and bone breadths. *Obesity, 13,* 626–634.

Gary, A., Campbell-Ruggaard, J., Goodheart, K. L., & Clopton, J. R. (2012). The physiology of anorexia nervosa. In K. L. Goodheart, J. R. Clopton, & J. J. Robert-McComb (Eds.), *Eating disorders in women and children* (pp. 47–59). Boca Raton: CRC Press.

Gelenberg, A. J., Wojcik, J. D., Gibson, C. J., & Wurtman, R. J. (1983). Tyrosine for depression. *Journal of Psychiatric Research, 17,* 175–180.

Gelenberg, A. J., Wojcik, J., Falk, W. E., Bellinghausen, B., & Joseph, A. B. (1989). CDP-choline for the treatment of tardive dyskinesia: A small negative series. *Comprehensive Psychiatry, 30,* 1–4.

Gelenberg, A. J., Wojcik, J., Falk, W. E., Baldessarini, R. J., Zeisel, S. H., Schoenfeld, D., & Mok, G. S. (1990). Tyrosine for depression: A double-blind trial. *Journal of Affective Disorders, 19,* 125–132.

Georgieff, M. K. (2007). Nutrition and the developing brain: Nutrient priorities and measurement. *American Journal of Clinical Nutrition, 85,* 614S–620S.

Gibson, R. S., Vanderkooy, P. D. S., MacDonald, A. C., Goldman, A., Ryan, B. A., & Berry, M. (1989). Growth limiting mild zinc-deficiency syndrome in some southern Ontario boys with low height percentiles. *American Journal of Clinical Nutrition, 49,* 1266–1273.

Gilbert, S. F. (2013). *Developmental biology* (10th ed.). Sunderland: Sinauer Associates.

Gleaves, D. H., & Latner, J. D. (2008). Evidence-based therapies for children and adolescents with eating disorders. In R. G. Steele, T. D. Elkin, & M. C. Roberts (Eds.), *Handbook of evidence-based therapies for children and adolescents: Bridging science and practice* (pp. 335–353). New York: Springer.

Gleaves, D. H., Miller, K. J., Williams, T. L., & Summers, S. A. (2000). Eating disorders: An overview. In K. J. Miller & J. S. Mizes (Eds.), *Comparative treatments for eating disorders*. New York: Springer.

Goldberg, I. K. (1980). L-tyrosine in depression. *The Lancet, 2*, 364.

Goldsmith, T. (2006). *Bulimia: Binging and purging*. Physch. Central. Available at: http://psychcentral.com/lib/bulimia-binging-and-purging/000283. Accessed 25 Mar 2015.

Goodpaster, B. H. (2002). Measuring body fat distribution and content in humans. *Current Opinion in Clinical Nutrition & Metabolic Care, 5*, 481–487.

Gopalan, C., & Naidu, A. N. (1972). Nutrition and fertility. *The Lancet, 300*, 1077–1079.

Gordon, R. A. (2000). *Eating disorders: Anatomy of a social epidemic*. Malden: Blackwell Publishers.

Gore, S. A., Vander Wal, J. S., & Thelen, M. H. (2001). Treatment of eating disorders in children and adolescents. In J. K. Thompson & L. Smolak (Eds.), *Body image, eating disorders, and obesity in youth: Assessment, prevention, and treatment* (pp. 293–311). Washington, DC: American Psychological Association.

Gorman, K. S. (1995). Malnutrition and cognitive development: Evidence from experimental/quasi-experimental studies among the mild-to-moderately malnourished. *Journal of Nutrition, 125*, 2239S–2244S.

Grant, J. E., Kim, S. W., & Eckert, E. D. (2002). Body dysmorphic disorder in patients with anorexia nervosa: Prevalence, clinical features, and delusionality of body image. *International Journal of Eating Disorders, 32*, 291–300.

Granthum-McGregor, S. (1995). A review of studies of the effect of severe malnutrition on mental development. *Journal of Nutrition, 125*, 2233S–2238S.

Granthum-McGregor, S. M., Fernald, L. C. H., Kagawa, R. M. C., & Walker, S. (2014). Effects of integrated child development and nutrition interventions on child development and nutritional status. In M. M. Black & K. G. Dewey (Eds.), *Every child's potential: Integrating nutrition and early childhood development interventions* (Annals of the New York Academy of Sciences) (Vol. 1308, pp. 11–32).

Greene, L. (1994). A retrospective view of iodine deficiency, brain development and behavior from studies in Equador. In J. Stanbury (Ed.), *The damaged brain of iodine deficiency*. New York: Cognizant Communication.

Grey, A., & Bolland, M. (2014). Research letter: Clinical trial evidence and use of fish oil supplements. *JAMA Internal Medicine, 174*, 460–462.

Gronbaek, M. (2004). Epidemiologic evidence for the cardioprotective effects associated with consumption of alcoholic beverages. *Pathophysiology, 10*, 83–92.

Growdon, J. H., Hirsch, M. J., Wurtman, R. J., & Wiener, W. (1977). Oral choline administration to patients with tardive dyskinesia. *New England Journal of Medicine, 297*, 524–527.

Grundman, M., & Delaney, P. (2002). Antioxidant strategies for Alzheimer's disease. *Proceedings of the Nutrition Society, 61*, 191–202.

Guo, X., Park, Y., Freedman, N. D., Sinha, R., & Hollenbeck, A. R. (2014). Sweetened beverages, coffee, and tea and depression risk among older US adults. *PLoS ONE, 9*, e94715. https://doi.org/10.1371/journal.pone.0094715.

Haag, M. (2003). Essential fatty acids and the brain. *Canadian Journal of Psychiatry, 48*, 195–203.

Hackman, D. A., & Farah, M. J. (2009). Socioeconomic status and the developing brain. *Trends in Cognitive Sciences, 13*, 65–73.

Hakkarainen, R., Partonen, T., Haukka, J., Virtamo, J., Albanes, D., & Lonnqvist, J. (2003). Association of dietary amino acids with low mood. *Depression and Anxiety, 18*, 89–94.

Hall, J. E., & Guyton, A. C. (2006). *Textbook of medical physiology*. St Louis: Elsevier Saunders.

Halterman, J., Kaczorowski, J., Aligne, C., Auinger, P., & Szilagyi, P. (2001). Iron deficiency and cognitive achievement among school-aged children and adolescents in the United States. *Pediatrics, 107*, 1381–1386.

Hannun, Y. A., & Bell, R. M. (1989). Functions of sphingolipids and sphingolipid breakdown products in cellular regulation. *Science, 243*, 500–507.

Harland, B. F. (2000). Caffeine and nutrition. *Nutrition, 1*, 522–526.

Hartman, E. (1987). The effect of L-tryptophan on the sleep-dream cycle in man. *Psychonomic science, 8*, 479–480.

Hebebrand, J., Exner, C., Hebebrand, K., Hotkamp, C., & Casper, H. (2003). Hyperactivity in patients with anorexia nervosa and in semi starved rats: Evidence for a pivotal role in hypoleptinemia. *Physiology & Behavior, 79*, 25–37.

Heinberger, D. C. (2014). Clinical manifestations of nutrient deficiencies and toxicities. In A. C. Ross, B. Caballero, R. J. Cousins, K. L. Tucker, & T. R. Ziegler (Eds.), *Modern nutrition in health and disease* (11th ed., pp. 757–770). Philadelphia: Lippincott Williams and Wilkins.

Helland, I. B., Smith, L., Saugstad, O. D., & Drevon, C. A. (2003). Maternal supplementation with very – long - chain n-3 fatty acids during pregnancy and lactation augments children's IQ at 4 years of age. *Pediatrics, 111*(1), e39–e44.

Helmchen, L. A., & Henderson, R. M. (2004). Changes in the distribution of body mass index of white US men, 1890–2000. *Annals of Human Biology, 31*, 174–181.

Henkin, R., Martin, B., & Agarwal, R. (1999). Decreased parotid saliva gustin/carbonic anhydrase VI secretion: An enzyme disorder manifested by gustatory and olfactory dysfunction. *American Journal of Medical Sciences, 318*, 380–391.

Hill, K., & Pomeroy, C. (2001). Assessment of physical status of children and adolescents with eating disorders and obesity. In J. K. Thompson & L. Smolak (Eds.), *Body image, eating disorders and obesity in youth: Assessment, prevention, and treatment* (pp. 171–191). Washington, DC: American Psychological Association.

Hill, J. O., Wyatt, H. R., Phelan, S., & Wing, R. R. (2005). The national weight control registry: Is it useful in helping deal with our obesity epidemic? *Journal of Nutrition Education and Behavior, 37*, 206–210.

Hirsch, M. J., & Wurtman, R. J. (1978). Lecithin consumption elevates acetylcholine concentration in rat brain and adrenal gland. *Science, 202*, 223–225.

Hoang, M. T., DeFina, L. F., & Willis, B. L. (2011). Association between low serum 25-hydroxyvitamin D and depression in a large sample of healthy adults: The cooper center longitudinal study. *Mayo Clinic Proceedings, 86*, 1050–1055.

Hoefer, C., & Hardy, M. C. (1929). Later development of breast fed and artificially fed infants: Comparison of physical and mental growth. *The Lancet, 92*(8), 615–619.

Hoek, H. W. (2006). Incidence, prevalence and mortality of anorexia nervosa and other eating disorders. *Current Opinion in Psychiatry, 19*, 389–394.

Hoes, M. L., & Curtis, B. (2012). Pharmaceutical approaches for treating eating disorders. In K. L. Goodheart, J. R. Clopton, & J. J. Robert-McComb (Eds.), *Eating disorders in women and children* (pp. 415–429). Boca Raton: CRC Press.

Horowitz, L. M. (2004). *Interpersonal foundations of psychopathology*. Washington, DC: American Psychological Association.

Hozl, J., & Godau, P. (1989). Receptor binding studies and *Valeriana officinalis* on the benzodiazepine receptor. *Planta Medica, 55*, 642.

Hrobjartsson, A., & Gotzche, P. C. (2001). Is the placebo powerless? An analysis of clinical trials comparing placebo with no treatment. *The New England Journal of Medicine, 344*(21), 1594–1602.

Hu, F. B., Li, T. Y., Colditz, G. A., Willsett, W. C., & Manson, J. E. (2003). Television watching and other sedentary behaviors in relation to risk of obesity and type-2 diabetes mellitus in women. *Journal of the American Medical Association, 289*, 1785–1791.

Hudson, C., Hudson, S. P., Hecht, T., & Mackenzie, J. (2005). Protein source tryptophan versus pharmaceutical grade tryptophan as an efficacious treatment for chronic insomnia. *Nutritional Neuroscience, 8*, 121–127.

Hypericum Depression Trial Study Group. (2002). Effect of *Hypericum perforatum* (St John's wort) in major depressive disorder: A randomized controlled trial. *Journal of the American Medical Association, 287,* 1807–1814.

Idjradinata, P., & Pollitt, E. (1993). Reversal of developmental delays in iron-deficient anemic infants treated with iron. *The Lancet, 34,* 1–4.

Innis, S. M. (1997). Polyunsaturated fatty acid nutrition in infants born at term. In J. Dobbing (Ed.), *Developing brain and behavior: The role of lipids in infants formula.* San Diego: Academic.

Innis, S. M., Nelson, C. M., Lwanga, D., Rioux, F. M., & Waslen, P. (1996). Feeding formula without arachidonic acid and docosahexaenoic acid has no effect on preferential looking acuity or recognition memory in healthy full term infants at 9 months of age. *American Journal of Clinical Nutrition, 64,* 40–46.

Insel, P., Turner, R. E., & Ross, D. (2004). *Nutrition* (2nd ed.). Sidbury: Jones and Bartlett Publishers.

Insel, P., Ross, D., McMahon, K., & Bernstein, M. (2013). *Discovering nutrition* (4th ed.). Burlington: Jones and Bartlett Learning.

Institute of Medicine. (1996). Division of biobehavioral sciences and mental disorders. Committee to study fetal alcohol syndrome. In K. Stratton, C. Howe, & F. Battaglia (Eds.), *Fetal alcohol syndrome: Diagnosis epidemiology, prevention, and treatment.* Washington, DC: National Academy Press.

Izzo, A. (2005). Herb-drug interactions: An overview of the critical evidence. *Fundamentals of Clinical Pharmacology, 19,* 1–16.

Jackson, I., Nuttall, A., & Perez-Cruet, J. (1979). Treatment of tardive dyskinesia with lecithin. *American Journal of Psychiatry, 136,* 1458–1459.

Jain, A. (2004). *What works for obesity? A summary of the research behind obesity interventions.* London: BMJ Publishing Group.

Jakicic, J. M., Clark, K., Coleman, E., Donnelly, J. E., & Foreyt, J. (2001). American college of sports medicine position stand: Appropriate intervention strategies for weight loss and prevention of weight regain for adults. *Medicine and Science in Sports and Exercise., 33,* 2145–2156.

James, J. E. (1997). *Understanding caffeine: A biobehavioral analysis.* Thousand Oaks: Sage.

James, J. E. (2004). Critical review of dietary caffeine and blood pressure: A relationship that should be taken more seriously. *Psychosomatic Medicine, 66,* 63–71.

James, W. P. T., Nelson, M., Ralph, A., & Leather, S. (1997). Socioeconomic determinants of health: The contribution of nutrition to inequalities of health. *British Medical Journal., 314,* 1545–1549.

Janowsky, D. S., & Overstreet, D. H. (1998). Acetylcholine. In P. J. Goodnick (Ed.), *Mania: Clinical and research perspectives* (pp. 135–155). Washington, DC: American Psychiatric Press, Inc.

Johansson, G., Wikman, A., Ahren, A. M., Hallmans, G., & Johansson, I. (2001). Underreporting of energy intake in repeated 24-hour recalls related to gender, age, weight status, day of interview, educational level, reported food intake, smoking habits and area of living. *Public Health Nutrition, 4,* 919–927.

Johnson, R. K., Appel, L. J., Brands, M., Howard, B. V., & Lefevre, M. (2009). Dietary sugars intake and cardiovascular health: A scientific statement from the American Heart Association. *Circulation, 120,* 1011–1020.

Johnson-Kozlow, M., Kritz-Silverman, D., Barrett-Connor, E., & Morton, D. (2002). Coffee consumption and cognitive function among older adults. *American Journal of Epidemiology, 156,* 842–850.

Jones, K. L., & Smith, D. W. (1973). Recognition of the fetal alcohol syndrome in early infancy. *The Lancet, 302,* 999–1001.

Jones, K. L., Smith, D. W., Ulleland, C. N., & Streissguth, A. P. (1973). Pattern of malinformation in offspring of chronic alcoholic mother. *The Lancet, 301,* 1267–1271.

Jouvet, M. (1968). Insomnia and decrease of cerebral 5-hydroxytryptamine after destruction of the raphe system in the cat. *Advances in Pharmacology, 6,* 265–279.

Juliano, L. M., & Griffiths, R. R. (2004). A critical review of caffeine withdrawal: Empirical validation of symptoms and signs, incidence, severity, and associated features. *Psychopharmacology, 176*, 1–29.

Kanarek, R. B. (1987). Neuropharmacological approaches to studying diet selection. In S. Kaufman (Ed.), *Amino acids in health and disease: New perspectives* (pp. 383–401). New York: Alan R. Liss.

Kanarek, R. B., & Carrington, C. (2004). Sucrose consumption enhances nicotine-induced analgesia in male and female smokers. *Psychopharmacology, 173*, 56–63.

Kanarek, R. B., Mandillo, S., & Wiatr, C. (2001). Chronic sucrose intake augments antinociception induced by injections of mu but not kappa opioid receptor agonists into the periaqueductal gray matter in male and female rats. *Brain Research, 920*, 97–105.

Kanarek, R. B., D'Anci, K. E., Mathes, W. F., Yamamoto, R., Coy, R. T., & Leibovici, M. (2005). Dietary modulation of the behavioral consequences of psychoactive drugs. In H. R. Lieberman, R. B. Kanarek, & C. Prasad (Eds.), *Nutritional neuroscience* (pp. 187–206). New York: CRC Press.

Kaplan, A. S., & Howlett, A. (2010). Pharmacotherapy for anorexia nervosa. In C. M. Grilo & J. E. Mitchell (Eds.), *The treatment of eating disorders: A clinical handbook* (pp. 175–186). New York: Guilford Press.

Kaplan, J. R., Shively, C. A., Botchin, M. B., Morgan, T. M., Howell, S. M., Manuck, S. B., Muldoon, M. F., & Mann, J. J. (1994). Demonstration of an association among dietary cholesterol, central serotonergic activity, and social behavior in monkeys. *Psychosomatic Medicine, 56*, 479–484.

Kaplan, J. R., Manuck, S. B., Fontenot, M. B., Muldoon, M. F., Shively, C. A., & Mann, J. J. (1997). The cholesterol-serotonin hypothesis: Interrelationships among dietary lipids, central serotonergic activity, and social behavior in monkeys. In M. Hillbrand & R. T. Spitz (Eds.), *Lipids, health, and behavior*. Washington, DC: American Psychological Association.

Katme, A. M. (1995). Analgesic effects of sucrose were known to the prophet. *British Medical Journal, 311*, 1169.

Kavey, R. E. (2010). How sweet it is: Sugar-sweetened beverage consumption, obesity, and cardiovascular risk in childhood. *Journal of the American Dietetic Association, 110*, 1450–1460.

Kellog, J. H. (1888). *Plain facts for old and young*. Burlington: Segner.

Kellog, J. H. (1919). *The itinerary of breakfast*. New York: Funk & Wagnalls.

Keys, A. J., Brozek, J., Henschel, A., Mickelson, O., & Taylor, H. L. (1950). *The biology of human starvation* (Vol. 2 vols). Minneapolis: University of Minnesota Press.

King, K., & Cousins, R. J. (2014). Zinc. In A. C. Ross, B. Caballero, R. J. Cousins, K. L. Tucker, & T. R. Ziegler (Eds.), *Modern nutrition in health and disease* (11th ed., pp. 189–205). Philadelphia: Lippincott Williams and Wilkins.

Kirk, S. F. L., Penney, T. L., & McHugh, T. L. F. (2010). Characterizing the obesogenic environment: The state of the evidence with directions for future research. *Obesity Reviews, 11*, 109–117.

Kirksey, A., Wachs, T., Yunis, F., Srinath, U., Rahmanifar, A., McCabe, G., Galal, O., Harrison, G., & Jerome, N. (1994). Relation of maternal zinc nutriture to pregnancy outcome and infant development in an Egyptian village. *American Journal of Clinical Nutrition, 60*, 782–792.

Klein, C. J. (Ed.). (2002). Nutrient requirements for preterm infant formulas. *The Journal of Nutrition, 132*(6), Suppl.1. 1395S-577S.

Kluck, A. S., Garos, S., & Johnson, L. (2012). Sexuality and eating disorders. In K. L. Goodheart, J. R. Clopton, & J. J. Roberet-McComb (Eds.), *Eating disorders in women and children* (pp. 181–194). Boca Raton: CRC Press.

Knutson, K. L., Spiegel, K., Penev, P., & van Cauter, E. (2007). The metabolic consequences of sleep deprivation. *Sleep Medicine Reviews, 11*, 163–178.

Kompoliti, K., & Horn, S. S. (2007). Drug-induced and iatrogenic neurological disorders. In C. G. Goetz (Ed.), *Textbook of clinical neurology* (pp. 1285–1318). Philadelphia: Saunders Elsevier.

Koren, G., Nulman, I., Chudley, A. E., & Loocke, C. (2003). Fetal alcohol spectrum disorder. *Canadian Medical Association Journal, 169*, 1181–1185.

Korkman, M., Kettunen, S., & Autti-Ramo, I. (2003). Neurocognitive impairment in early adolescence following prenatal alcohol exposure of varying duration. *Child Neuropsychology, 9*, 117–128.

Kotani, S., Sakaguchi, E., Warashina, S., Matsukawa, N., & Ishikura, Y. (2006). Dietary supplementation of arachidonic and docosahexaenoic acids improves cognitive dysfunction. *Neuroscience research, 56*, 159–164.

Kramer, M. S., Aboud, F., Mironova, E., Vanilovich, I., Plat, R. W., & Matusch, L. (2008). Breastfeeding and child cognitive development: New evidence from a large randomized trial. *Archives of General Psychiatry, 65*, 578–584.

Krondl, M. (1990). Conceptual models. In G. H. Anderson (Ed.), *Diet and behavior: Multidisciplinary approaches* (pp. 5–15). London: Springer.

Kruesi, M. J. P., Rapport, J. L., Hamburger, S., Hibbs, E., & Potter, W. Z. (1990). Cerebrospinal fluid monoamine metabolites, aggression, and impulsivity in disruptive behavior disorders of children and adolescents. *Archives of General Psychiatry, 47*, 419–442.

Krugman, S. D., & Dubowitz, H. (2003). Failure to thrive. *American Family Physician, 68*, 879–886.

Kunhn, W., Roebroek, R., Blom, H., van Oppenraaij, D., Przuntek, H., Kretschmer, A., Buttner, T., Woitalla, D., & Muller, T. (1998). Elevated plasma levels of homocysteine in Parkinson's didease. *European Neurology, 40*, 225–227.

Kwiterovich, P. O. (1986). Biochemical, clinical, epidemiologic, genetic, and pathologic data in the pediatric age group relevant to the cholesterol hypothesis. *Pediatrics, 78*, 349–362.

Lauberg, P. (2014). Iodine. In A. C. Ross, B. Caballero, R. J. Cousins, K. L. Tucker, & T. R. Ziegler (Eds.), *Modern nutrition in health and disease* (11th ed., pp. 217–224). Philadelphia: Lippincott Williams and Wilkins.

Laurent, D., Schneider, K. E., Prusacyck, W. K., Franklin, C., & Vogel, S. M. (2000). Effect of caffeine on muscle glycogen utilization and the neuroendocrine axis during exercise. *Journal of Clinical Endocrinology and Metabolism, 85*, 2170–2175.

Le Boeuf, R. (2003). Homocysteine and Alzheimer's disease. *Journal of the American Dietetic Association, 103*, 304–307.

Leatherwood, P. D., & Pollet, P. (1983). Diet-induced mood changes in normal populations. *Journal of Psychiatric Research, 17*(2), 147–154.

Leibenluft, E., Fiero, P., Bartko, J. J., Moul, D. E., & Rosenthal, N. E. (1993). Depressive symptoms and the self-reported use of alcohol, caffeine, and carbohydrates in normal volunteers and four groups of psychiatric outpatients. *American Journal of Psychiatry, 150*, 294–301.

Lemoine, P., Haronsseau, H., Borteyu, J. P., & Menuet, J. C. (1968). Les enfants de patents alcoolique observes a propose de 127 cas. *Qust Medcical, 25*, 476–482.

Lester, B. M., Klein, R. E., & Martinez, S. J. (1975). The use of habituation in the study of the effects of infantile malnutrition. *Developmental Psychobiology, 8*, 541–546.

Levander, O. A., & Burk, R. F. (1992). Selenium. In M. L. Brown (Ed.), *Present knowledge in nutrition* (6th ed., pp. 268–273). Washington, DC: International Life Science Institute-Nutrition Foundation.

Levine, J. A. (2004). Non-exercise activity thermogenesis (NEAT). *Nutrition Reviews, 62*, S82–S97.

Levine, J. A., Eberhardt, N. L., & Jensen, M. D. (1999). Role of non exercise activity thermogenesis in resistance in fat gain in humans. *Science, 283*, 212–214.

Levine, J. A., Vander Weg, M. W., Hill, J. O., & Klesges, R. C. (2006). Non-exercise activity thermogensesis: The crouching tiger hidden dragon societal weight gain. *Arteriosclerosis, Thrombosis, and Vascular Biology, 26*, 729–736.

Lieberman, H. R. (2001). The effects of ginseng, ephedrine, and caffeine on cognitive performance, mood and energy. *Nutrition Reviews, 59*, 91–102.

Lieberman, H. (2003). Nutrition, brain function and cognitive performance. *Appetite, 40*, 245–254.

Lieberman, H. R., Corkin, S., Spring, B. J., Growdon, J. H., & Wurtman, R. J. (1983). Mood, performance, and pain sensitivity: Changes induced by food constituents. *Journal of Psychiatric Research, 1983*, 135–145.

Lopez-Figueroa, A. L., Norton, C. S., Lopez-Figueroa, M. O., Armelini-Dodel, D., & Burke, S. (2004). Serotonin 5-HT1A, 5-HT1B, and 5-HT2A receptor mRNA expression in subjects with major depression, bipolar disorder and schizophrenia. *Biological Psychiatry, 55*, 225–233.

Lorist, M. M., & Tops, M. (2003). Caffeine, fatigue, and cognition. *Brain and Cognition, 53*, 82–94.

Lozoff, B., Klein, N., Nelson, E., McClish, D., Manuel, M., & Chacon, M. (1998). Behavior of infants with iron deficiency anemia. *Child Development, 69*, 24–36.

Lozoff, B., Jimenez, E., Hagen, J., Mollen, E., & Wolf, A. (2000). Poorer behavioral and developmental outcome more than 10 years after treatment for iron deficiency in infancy. *Pediatrics, 105*, e51.

Lozoff, B., Jimenez, E., & Smith, J. B. (2006). Double burden of iron deficiency in infancy and low socioeconomic status: A longitudinal analysis of cognitive test scores to age 19 years. *Archives of Pediatrics & Adolescent Medicine, 160*, 1108–1113.

Lucassen, E. A., Rother, K. I., & Gizza, G. (2012). Interacting epidemics? Sleep curtailment, insulin resistance, and obesity. *Annals of the New York Academy of Sciences, 1264*, 110–134.

Lucca, A., Lucini, V., Piatti, E., Ronchi, P., & Smeraldi, E. (1992). Plasma tryptophan levels and plasma tryptophan/neutral amino acids ratio in patients with mood disorder, patients with obsessive-compulsive disorder and normal subjects. *Psychiatric Research, 44*, 85–91.

Luttinger, N., & Dicum, G. (2006). *The coffee book: Anatomy of an industry from crop to the last drop.* New York: The New Press.

Lysaght, P. (1987). "When i makes tea, i makes tea": The case of tea in Ireland. *Ulster Folklife, 33*, 48–49.

Macdiarmid, J. I., & Hetherington, M. M. (1995). Mood modulation by food: An exploration of affect and cravings in "chocolate addicts". *British Journal of Clinical Psychology, 34*, 129–138.

Magazine Monitor. (2014). *Why do Britons drink so much instant coffee?* Available at: http://www.bbc.com/news/blogs-magazine-monitor-26869244. Accessed 25 Mar 2015.

Magill, R. A., Waters, W. F., Bray, G. A., Volaufova, J., & Smith, S. R. (2003). Effects of tyrosine, phentermine, caffeine D-amphetamine, and placebo on cognitive and motor performance deficits during sleep deprivation. *Nutritional Neurosciences, 6*, 237–246.

Mailot, F., Farad, S., & Lamisse, F. (2001). Alcohol and nutrition. *Pathologie Biologie, 49*, 683–688.

Manco, M. (2013). Bariatric surgery and reversal of metabolic disorders. In D. Bagchi & H. C. Preuss (Eds.), *Obesity: Epidemiology, pathophysiology, and prevention* (2nd ed., pp. 931–945). Boca Raton: CRC Press.

Mandel, H. G. (2002). Update on caffeine consumption, disposition, and action. *Food and Chemical Toxicology, 40*, 1231–1234.

Marketdata. (2014). *The US weight loss market: 2014 status report & forecast.* Marketdata Enterprises Inc. Available at: http://www.marketresearch.com/Marketdata-Enterprises-Inc-v416/Weight-Loss-Status-Forecast-8016030. Accessed 25 Mar 2015.

Marotta, R. B., & Floch, M. H. (1991). Diet and nutrition in ulcer disease. *The Medical Clinics of North America, 75*, 967–979.

Mason, R. P., Rubin, R. T., Mason, P. E., & Tulenko, T. N. (1997). Molecular mechanisms underlying the effects of cholesterol on neuronal cell membrane function and drug-membrane interactions. In M. Hillbrand & R. T. Spitz (Eds.), *Lipids, health, and behavior.* Washington, DC: American Psychological Association.

Maughan, R. J., & Griffin, J. (2003). Caffeine ingestion and fluid balance: A review. *Journal of Human Nutrition and Dietetics, 16*, 141–420.

Mayeno, A. N., & Gleich, G. J. (1994). Eosinophilia-myalgia syndrome and tryptophan production: A cautionary tale. *Trends in Biotechnology, 12*, 346–352.

McEachin, R. C., Keller, B. J., Saunders, E. F., & McInnis, M. G. (2008). Modeling gene-by-environment interaction in comorbid depression with alcohol use disorders via an integrated bio-informatics approach. *BioData Mining, 1*, 2.

McElroy, S. L., Kotwal, R., Malhotta, S., Nelson, E. B., Keck, P. E., Jr., & Nemeroff, C. B. (2004). Are mood disorders and obesity related? A review for the mental health professional. *Journal of Clinical Psychiatry, 65*, 634–651.

McLaren, L. (2007). Socioeconomic status and obesity. *Epidemiologic Reviews, 29*, 29–48.

Melis, M., Atzori, E., Cabras, S., Zonza, A., & Calto, M. C. (2013). A gustin gene polymorphism as a mechanistic link between PROP tasting and taste papilla density and morphology. *PLoS One, 8*, e74151.

Menkes, D. B., Coates, D. C., & Fawcett, J. P. (1994). Acute tryptophan depletion aggravates premenstrual syndrome. *Journal of Affective Disorders, 32*(1), 37–44.

Mercer, M. E., & Holder, M. D. (1997). Antinociceptive effects of palatable sweet ingesta on human responsivity to pressure pain. *Physiology & Behavior, 61*, 311–318.

Miller, M. A., & Cappuccio, F. P. (2013). Sleep and obesity. In D. Bagchi & H. C. Preuss (Eds.), *Obesity: Epidemiology, pathophysiology, prevention* (2nd ed., pp. 179–190). Boca Raton: CRC Press.

Mitchell, J. E., & Crow, S. J. (2010). In W. S. Agras (Ed.), *Medical comorbidities of eating disorders* (pp. 259–266). New York: The Oxford University Press.

Miyashita, K., & Hosokawa, M. (2013). Antiobesity by marine lipids. In D. Bagchi & H. C. Preuss (Eds.), *Obesity: Epidemiology, pathophysiology, and prevention* (2nd ed., pp. 651–665). Boca Raton: CRC Press.

Mokdad, A. H., Marks, J. S., Stroup, D. F., & Gerberding, J. L. (2004). Actual causes of death in the United States. *Journal of the American Medical Association, 291*, 1238–1245.

Mortensen, E. L., Michaelsen, K. F., Sanders, S. A., & Reinisch, J. M. (2002). The association between duration of breastfeeding and adult intelligence. *Journal of the American Medical Association, 287*(18), 2365–2371.

Morton, N. M., Paterson, J. M., Masuzaki, H., Holmes, M. C., & Staels, B. (2004). Novel adipose tissue-mediated resistance to diet-induced visceral obesity in 11 β–hydroxysteroid dehydrogenase Type-1 deficient mice. *Diabetes, 53*, 931–938.

Mourao, D. M., Bressan, J., Campbell, W. W., & Mattes, R. D. (2007). Effects of food form on appetite and energy intake in lean and obese young adults. *International Journal of Obesity, 31*, 1688–1695.

Murphy, J. M., Wehler, C. A., Pagano, M. E., Little, M., Kleinman, R. E., & Jellinek, M. S. (1998). Relationship between hunger and psychosocial functioning in low-income American children. *Journal of the American Academy of Child & Adolescent Psychiatry, 37*, 163–170.

Nair, K. P. P. (1996). The buffering power of plant nutrients and effects on availability. *Advances in Agronomy, 57*, 237–287.

Nair, K. P. P. (2010). *The agronomy and economy of important tree crops of the developing world*. London/Burlington: Elsevier.

Nair, K. P. P. (2011). *Agronomy and economy of black pepper and cardamom*. London/Waltham: Elsevier.

Nair, K. P. P. (2013a). *The agronomy and economy of turmeric and ginger*. London/Waltham: Elsevier.

Nair, K. P. P. (2013b). The buffer power concept and its relevance in African and Asian soils. *Advances in Agronomy, 121*, 447–516.

Nair, K. P. P. (2016). *The nutrient buffer power concept for sustainable agriculture* (p. 434). Chennai: Notion Press.

Naismith, D. J. (1969). The foetus as parasite. *Proceedings of the Nutrition Society, 28*, 25–31.

Nakashima, Y., & Suzue, R. (1982). Effect of nicotinic acid on myelin lipids in brain of the developing rat. *Journal of Nutritional Science and Vitaminology, Tokyo, 28*, 491–500.

Nathan, P. J., Baker, A., Carr, E., Earle, J., Jones, M., Nieciecki, M., Hutchison, C., & Stough, C. (2001). Cholinergic modulation of cognitive function in healthy subjects: Acute effects of donepezil, a cholinesterase inhibitor. *Human Psychopharmacology, 16*(6), 481–483.

National Coffee Association. (2013). *2013 national coffee drinking trends.* Available at: http://www.ncausa.Org/i4a/pages/index.cfm?page1D=924. Accessed 25 Mar 2015.

National Institute on Alcohol Abuse and Alcoholism. (2014). *Alcohol facts and statistics.* Available at: http://www.niaaa.nih.ov/alcohol-health/overview-alcohol-consumption/alcohol-facts-and-statistics. Accessed 25 Mar 2015.

Nawrot, P., Jordon, S., Eastwood, J., Rotstein, J., Hugenholtz, A., & Feeley, M. (2003). Effects of caffeine on human health. *Food Additives and Contaminants, 20*, 1–30.

Nederkroon, C., van Eijs, Y., & Jansen, A. (2004). Restrained eaters act on impulse. *Personality and Individual Differences, 37*, 1651–1658.

Nelson, E. C., Heath, A. C., Buchholz, K. K., Madden, P. A. F., & Fu, Q. (2004). Genetic epidemiology of alcohol-induced blackouts. *Archives of general Psychiatry, 61*, 257–263.

Nestle, M., & Nesheim, M. (2012). *Why calories count: From science to politics.* Berkley: University of California Press.

Newman, J. I. (2000). Wine. In K. F. Kiple & K. C. Omelas (Eds.), *The Cambridge world history of food* (pp. 730–737). Cambridge: Cambridge University Press.

Ng, M., Fleming, T., Robinson, M., Thomson, B., & Graetz, N. (2014). Global, national and regional prevalence of overweight and obesity in children and adults during 1980–2013: A systematic analysis for the Global Burden of Disease Study 2013. *The Lancet, 384*, 766–781.

Nicholas, D. E., & Nicholas, C. D. (2008). Serotonin receptors. *Chemical Reviews, 108*, 1614–1641.

Nieman, D. C. (2010). *Exercise testing and prescription* (7th ed.). New York: McGraw - Hill Higher Education.

Noakes, M., Keogh, J. B., Foster, P. R., & Clifton, P. M. (2005). Effect of an energy-restricted, high-protein, low-fat diet relative to a conventional high-carbohydrate, low-fat diet on weight loss, body composition, nutritional status, and markers of cardiovascular health in obese women. *American Journal of Nutrition, 81*, 1298–1306.

Noordzij, M., Uiterwaal, C., Arends, L. R., Kok, F. J., Grobbe, D. E., & Geleijnse, J. M. (2005). Blood pressure response to chronic intake of coffee and caffeine: a meta-analysis of randomized controlled trials. *Journal of Hypertension, 23*, 921–928.

O'Shea, R. S., Dasarathy, S., & McCullough, A. J. (2010). Alcoholic disease. *Hepatology, 51*, 307–328.

Oakley, G. P., Erckson, J. D., James, L. M., Mulinare, J., & Cordero, J. F. (1994). Prevention of folic acid preventable spina bifida and anencephaly. In G. Bock & J. Marsh (Eds.), *Neural tube defects* (pp. 212–222). Cheicester: Wiley.

Odutaga, A. A. (1982). Effects of low-zinc status and essential fatty acid deficiency on growth and lipid composition of rat brain. *Clinical Experimental Journal of Pharmacology and Physiology, 9*, 213–221.

Otte, C., Zhao, S., & Whooley, M. A. (2012). Statin use and risk of depression in patients with coronary heart disease: Longitudinal data from the Heart and Soul Study. *Journal of Clinical Psychiatry, 73*, 610–615.

Owasoyo, J. O., Neri, D. F., & Lamberth, J. G. (1992). Tyrosine and its potential use as a countermeasure to performance decrement in military sustained operations. *Aerospace Medical Association, 63*, 364–369.

Paddon-Jones, D., Westman, E., Mattes, R. D., Wolfe, R. R., Astrup, A., & Weterterp-Plantenga, M. (2008). Protein, weight management and satiety. *American Journal of Clinical Nutrition, 87*, 1558S–1561S.

Pardridge, W. M. (1986). Blood-brain barrier transport of nutrients. *Nutrition reviews, 44*, 15–25.

Pedersen, S. D., Sjodin, A., & Astrup, A. (2012). Obesity as a health risk. In J. W. Erdman, I. A. McDonald, & S. H. Zeisel (Eds.), *Present knowledge in nutrition* (10th ed., pp. 709–720). Ames: International Life Sciences Institute.

Peet, M., & Horrobin, D. F. (2002). A close-ranging study of the effects of ethyl-eicosapentaenoate in patients with ongoing depression despite apparently adequate treatment with standard drugs. *Archives of General Psychiatry, 59*, 913–919.

Peet, M., Brind, J., Ramchand, C. N., Shash, S., & Vankar, G. K. (2001). Two double-blind placebo-controlled pilot studies of eicosapentaenoic acid in the treatment of schizophrenia. *Schizophrenia Research, 49*, 243–251.

Perry, M. L., Gamallo, J. L., & Bernard, E. A. (1986). Effect of protein malnutrition on glycoprotein synthesis in rat cerebral cortex slices during the period of brain growth spurt. *Journal of Nutrition, 116*, 2486–2489.

Perry, L. A., Stigger, C. B., Aimnsworth, B. E., & Zhang, J. (2009). No association between cognitive achievements, academic performance, and serum cholesterol concentrations among school-aged children. *Nutritional Neuroscience, 12*, 160–166.

Petersen, R. C., Thomas, R. G., Grundman, M., Bennett, D., & Doody, R. (2005). Vitamin E and donepezil for the treatment of mild cognitive impairment. *New England Journal of Medicine, 352*, 2379–2388.

Petty, F., Kramer, G., Wilson, L., & Chae, Y. L. (1993). Learned helplessness and *in vivo* hippocampal norepinephrine release. *Pharmacology, Biochemistry and Behavior, 46*, 231–235.

Pfrieger, F. W. (2003). Role of cholesterol in synapse formation and function. *Biochemica et Biophysica Acta, 1610*, 271–280.

Pike, K. M., Loeb, K., & Vitousek, K. (2001). Cognitive-behavioral therapy for anorexia nervosa and bulimia nervosa. In J. K. Thompson (Ed.), *Body image, eating disorders, and obesity: An integrative guide for assessment and treatment* (pp. 253–302). Washington, DC: American Psychological Association.

Pi-Sunyer, F. X. (2007). Weight management in diabetes prevention. In R. F. Kushner & D. H. Bessesen (Eds.), *Treatment of the obese patient* (pp. 243–263). Totowa: Humana Press Inc..

Pittler, M. H., & Ernst, E. (2003). Kava extract for treating anxiety. *The Cochrane Database of Systematic Reviews, 1*, CD003383. https://doi.org/10.1002/14651858.CD003383.

Pogoto, S. L., Spring, B., McChargue, D., Hitsman, B., & Smith, M. (2009). Acute tryptophan depletion and sweet food consumption by overweight adults. *Eating Behaviors, 10*, 26–41.

Polivy, J., Zeitlin, S., Herman, C., & Beal, A. (1994). Food restriction and binge eating: A study of former prisoners of war. *Journal of American Psychology, 103*, 409–411.

Pomeroy, C. (1996). Anorexia nervosa, bulimia nervosa, and binge eating disorder assessment of physical status. In J. K. Thompson (Ed.), *Body image, eating disorders, and obesity* (pp. 177–203). Washington, DC: American Psychological Association.

Popkin, B. M., Adair, L. S., & Ng, S. W. (2012). Global nutrition transition and the pandemic of obesity in developing countries. *Nutrition reviews, 70*, 3–21.

Prasad, A. (1988). Clinical spectrum and diagnostic aspects of human zinc deficiency. In *Essential and toxic trace elements in human health and disease*. New York: Alan R. Liss.

Prince, M., Prina, M., & Guerchet, M. (2013). *Journey of caring: An analysis of long-term care for dementia* (World Alzheimer Report 2013). Available at: http://www.alz.co.uk/worldreport 2013. Accessed 24 Mar 2015.

Puder, J. J., & Munsch, S. (2010). Psychological correlates of childhood obesity. *International Journal of Obesity, 34*, S37–S43.

Qawasmi, A., Landeros-Weisenberger, A., Leckman, J. F., & Bloch, M. H. (2012). Meta-analysis of long-chain polyunsaturated acid supplementation of formula and infant cognition. *Pediatrics, 129*, 1141–1149.

Quetelet, L. A. J. (1871). *Anthropometrie ou mesure des differentes faculties de l'homme* [Anthropometry or measurement of different characteristics of man]. Brusels: Muqerdt

Rand, C. S., & MacGregor, A. M. (1991). Successful weight loss following obesity surgery and the perceived liability of morbid obesity. *International Journal of Obesity, 15*, 577–579.

Ray, O., & Ksir, C. (2013). *Drugs, society and human behavior* (15th ed.). New York: McGraw Hill.

Reddy, T. S., & Ramakrishnan, C. V. (1982). Effects of maternal thiamine deficiency on the lipid composition of rat whole brain, gray matter and white matter. *Neurochemistry International, 4*, 495–499.

Reid, M., & Hammersley, R. (1995). Effects of carbohydrate intake on subsequent food intake and mood state. *Physiology & Behavior, 58*, 421–427.

Reid, M., & Hammersley, R. (1999). The effects of sucrose and maize-oil on subsequent food intake and mood. *British Journal of Nutrition, 82*, 447–455.

Reisbick, S., Neuringer, M., Gohl, E., Waid, R., & Anderson, G. J. (1997). Visual attention in infant monkeys: Effects of dietary fatty acids and age. *Developmental Psychology, 33*, 387–395.

Repo-Tiihonen, E., Halonen, P., Tiihonen, J., & Virkkunen, M. (2002). Total serum cholesterol level, violent criminal offences, suicidal behavior, mortality and the appearance of conduct disorder in Finnish male criminal offenders with antisocial personality disorder. *European Archives of Psychiatry and Clinical Neuroscience, 252*, 8–11.

Ricciuti, H. (1993). Nutrition and mental development. *Current Directions in Psychological Science, 2*, 43–46.

Rice, A. L., Sacco, L., Hyder, A., & Black, R. E. (2000). Malnutrition as an underlying cause of childhood deaths associated with infectious diseases in developing countries. *Bulletin of the World health Organization, 78*, 1207–1221.

Robert-McComb, J. J., & McCullough, B. (2012). The physiology of bulimia nervosa. In K. L. Goodheart, J. L. Clopton, & J. J. Robert-McComb (Eds.), *Eating disorders in women and children* (pp. 61–74). Boca Raton: CRC Press.

Roberts, D. L., Dive, C., & Renehan, A. G. (2010). Biological mechanisms linking obesity and cancer risk: New perspectives. *Annual Review of Medicine, 61*, 310–316.

Rodin, J. (1990). Behavior: Its definition measurement in relation to dietary intake. In G. H. Anderson (Ed.), *Diet and behavior: Multidisciplinary approaches* (pp. 57–72). London: Springer.

Roehrs, T., & Roth, T. (2008). Caffeine: Sleep and daytime sleepiness. *Sleep Medicine Reviews, 12*, 153–162.

Rogers, P. J., Martin, J., Smith, C., Heatheley, S. V., & Smith, H. J. (2003). Absence of reinforcing, mood and psychomotor performance effects of caffeine in habitual non-consumers of caffeine. *Psychopharmacology, 167*, 545–562.

Rolls, B. J. (2003). The supersizing of America: Portion size and the obesity epidemic. *Nutrition Today, 38*, 42–53.

Rolls, B. J., Shide, D. J., Thorwart, M. L., & Ulbrecht, J. S. (1998). Sibutramine reduces food intake in non-dieting women with obesity. *Obesity Research, 6*, 1–11.

Rolls, B. J., Drewnowski, A., & Ledikwe, J. H. (2005). Changing the energy density of the diet as a strategy for weight management. *Journal of the American Dietetic Association, 105*, S98–S103.

Romano, S., Halmi, K., Sarkar, N., Koke, S., & Lee, J. (2002). A placebo-controlled study of fluoxetine in continued treatment of bulimia nervosa after successful acute fluoxetine treatment. *American Journal of Psychiatry. 74, 159*, 96–102.

Rosenberg, G. S., & Davis, K. L. (1982). The use of cholinergic precursors in neuropsychiatric diseases. *American Journal of Clinical Nutrition, 36*, 709–720.

Rosenzweig, M. R., & Leiman, A. L. (1989). *Physiological psychology*. New York: Random House.

Ruhe, H. G., Mason, N. S., & Schene, A. H. (2007). Mood is indirectly related to serotonin, norepinephrine and dopamine levels in humans: A meta-analysis of monamine depletion studies. *Molecular Psychiatry, 12*, 331–359.

Samanin, R., & Garattini, S. (1990). The pharmacology of serotonergic drugs affecting appetite. In R. J. Wurtman & J. J. Wurtman (Eds.), *Nutrition and the brain* (Vol. 8, pp. 163–192). New York: Raven Press.

Sandstead, H., Penland, J., Alcock, N., Dayal, H., Chen, X., Li, J., Zhao, F., & Yang, J. (1998). Effects of repletion with zinc and other micronutrients on neuropsychologic performance and growth of Chinese children. *American Journal of Clinical Nutrition, 68*, 470S–475S.

Sano, M., Ernesto, C., Thomas, R. G., Klauber, M. R., & Schafer, K. (1997). A controlled trial of selegiline, alpha-tocopherol, or both as treatment for Alzheimer's disease. The Alzheimer's disease cooperative study. *New England Journal of Medicine, 336*, 1216–1222.

Santos, V. G. F., Santos, V. R. F., Felippe, L. L., Almeida, J. W., & Bertuzzi, R. (2014). Caffeine reduces reaction time and improves performance in simulated –contest of Taekwondo. *Nutrients, 6*, 637–649.

Santuccis, P. (2014). *Glossary of treatment terms. A brief overview of therapies used in the treatment of eating disorders; A consumer's guide.* National Association of Anorexia Nervosa and Associated Disorders. Available at: http://www.anad.org/get-information/information-about-treatment. Accessed 25 Mar 2015.

Saris, W. H. M. (2001). Very-low –calorie diets and sustained weight loss. *Obesity, 9*, 295S–301S.

Sauceman, F. W. (2009). Dr *Enuf*: A new age neutraceutical with a patent medicine pedigree. In *The place setting* (pp. 89–97). Macon: Mercer University Press.

Saunders, E. E., Zhang, P., Copeland, J. N., McInnis, M. G., & Zollner, S. (2009). Suggestive linkages at 9p22 in bipolar disorder weighted by alcohol abuses. *American Journal of Molecular Genetics Part B: Neuropsychiatric Genetics, 150B*, 1133–1138.

Save the Children. (2012). *State of the world's mothers 2012: Nutrition in the first 1000 days.* Available at: http://www.Savethechildren.org/atf/cf%7B9def2ebe-10ae-432c-9bdO-df91d2eba74%7D/STATE-OF-THE WORLDS-MOTHERS-REPORT-2012-FINAL.PDF. Accessed 24 Mar 2015.

Schardt, D. (2003). Eat less and live longer? Does calorie restriction work? *Nutrition Action Health Letter, 30*(1), 3–6.

Schardt, D., & Schmidt, S. (1996). Caffeine: The inside scoop. *Nutrition Action Newsletter, 23*, 1–7.

Schreurs, B. G. (2010). The effect of cholesterol on learning and memory. *Neuroscience and Behavioral Reviews, 34*, 1366–1379.

Schuman, M., Gitlin, M. J., & Fairbanks, L. (1987). Sweets, chocolate and atypical depressive traits. *Journal of Nervous and Mental Disorders, 175*, 491–495.

Schurch, B., & Scrimshaw, N. S. (Eds.). (1990). *Activity, energy expenditure and energy requirements of infants and children.* Lausanne: Nestle Foundation.

Schwartz, J. H. (2001). *Neurotransmitters. Encyclopedia of life sciences.* Wiley Online Library. https://doi.org/10.1038/npg.els.0000287.

Schwartz, S. M., & Savastano, D. M. (2013). History and regulation of prescription and over-the-counter weight loss drugs. In D. Bagchi & H. C. Preuss (Eds.), *Obesity: Epidemiology, pathophysiology, and prevention* (2nd ed., pp. 313–348). Boca Raton: CRC Press.

Schwartz, M., Chambliss, H. O., Brownell, K., Blair, S., & Billington, C. (2003). Weight bias among health professionals specializing in obesity. *Obesity Research, 11*(9), 1033–1039.

Schweinsburg, B., Alhassoon, O., Taylor, M., Gonzalez, R., & Videen, J. S. (2003). Effects of alcoholism and gender on brain metabolism. *American Journal of Psychiatry, 160*, 1180–1183.

Sevick, M. A., Piraino, B., Sereika, S., Starrett, T., & Bender, C. (2005). A preliminary study of PDA-based dietary self monitoring in hemodialysis patients. *Journal of Renal Nutrition, 15*, 304–311.

Shukla, S. D., & Halenda, S. P. (1991). Phospholipase D in cell signaling and its relationship to phospholipase C. *Life Sciences, 48*, 851–866.

Simic, M., & Eisler, I. (2012). Family and multifamily therapy. In J. Fox & K. Goss (Eds.), *Eating and its disorders* (pp. 260–279). Malden: Wiley-Blackwell.

Sindler, A. J., Wellman, N. S., & Stier, O. B. (2004). Holocaust survivors report long-term effects on attitudes toward food. *Journal of Nutrition Education and Behavior, 36*, 189–196.

Sitaram, N., Weingartner, H., Caine, E. D., & Gillin, J. C. (1978). Choline: Selective enhancement of serial learning and encoding of low imagery words in man. *Life Sciences, 22*, 1555–1560.

Sjodin, A. M., Forslund, A. N., Westerterp, H. N., Andersson, A. B., Forslund, J. M., & Hambraaeus, L. M. (1996). The influence of physical activity on BMR. *Medicine and Science in Sports and Exercise, 28*, 85–91.

Smith, A. (2005). Caffeine. In H. Lieberman, R. Kanarek, & C. Prasad (Eds.), *Nutritional neuroscience* (pp. 341–361). New York: CRC Press.

Smith, M. J., & Garrett, R. H. (2005). A heretofore undisclosed crux of eosinophilia-myalgia syndrome: Compromised histamine degradation. *Inflammation Research, 54*, 435–450.

Smith, C., Klosterbuer, A., & Levine, A. S. (2009). Military experience strongly influences post-service eating behavior and BMI status in American veterans. *Appetite, 52*, 280–289.

Snitz, B. E., O'meara, E. S., Carlson, M. C., Arnold, A. M., & Ives, D. G. (2009). Ginkgo biloba for preventing cognitive decline in older adults: A randomized trial. *Journal of the American Medical Association, 302*, 2663–2670.

Sokol, R. J., Delaney-Black, V., & Nordstrom, B. (2003). Fetal alcohol spectrum disorder. *Journal of the American Medical Association, 290*, 2996–2999.

Solomon, A., Kivipelto, M., Wolozin, B., Zhou, J., & Whitmer, R. A. (2009a). Midlife serum cholesterol and increased risk of Alzheimer's and vascular dementia three decades later. *Dementia and Geriatric Cognitive Disorders, 28*, 75–80.

Solomon, A., Kareholt, I., Ngandu, T., Wolozin, B., & MacDonald, S. W. S. (2009b). Serum cholesterol, statins, and cognition in non-demented elderly. *Neurobiology of Aging, 30*, 1006–1009.

Spector, A. A., & Yorek, M. A. (1985). Membrane lipid composition and cellular function. *Journal of Lipid Research, 26*, 1015–1035.

Spedding, S. (2014). Vitamin D and depression: A systematic review and meta-analysis comparing studies with and without biological flaws. *Nutrients, 6*, 1501–1518.

Spina, D. (2003). Theophylline and PDE4 inhibitors in asthma. *Current Opinion in Pulmonary Medicine, 9*, 57–64.

Spring, B. (1986). Effects of foods and nutrients on the behavior of normal individuals. In R. J. Wurtman & J. J. Wurtman (Eds.), *Nutrition and the brain* (pp. 1–48). New York: Raven Press.

Spring, B., Chiodo, J., Harden, M., Bourgeois, M. J., Mason, J. D., & Lutherer, L. (1989). Psychobiological effects of carbohydrates. *Journal of Clinical Psychiatry, 50*(5), 27–33.

Stice, E., Wonderlich, S., & Wade, E. (2006). Eating disorders. In M. Hersen, J. C. Thomas, & R. T. Ammerman (Eds.), *Comprehensive handbook of personality and psychopathology* (pp. 330–347). Hoboken: Wiley.

Sturm, R. (2007). Increases in morbid obesity in the USA: 2000–2005. *Public Health, 121*, 492–496.

Su, K.-P., Huang, S.-H., Chiu, C.-C., & Shen, W. W. (2003). Omega-3 fatty acids in major depressive disorder: A preliminary double-blind, placebo-controlled trial. *European Neuropsychopharmacology, 13*, 267–271.

Subramanya, S. B., Subramanian, V. S., & Said, H. M. (2010). Chronic alcohol consumption and intestinal thiamin absorption: Effects on physiological and molecular parameters of the uptake process. *American Journal of Physiology: Gastrointestinal and Liver Physiology, 299*, G23–G31.

Sullivan, W. C. (1989). A note on the influence of maternal inebriety on the offspring. *Journal of Mental Science, 45*, 489–503.

Super, C. M., Herrara, M. G., & Mora, J. O. (1990). Long-term effects of food supplementation and psychosocial intervention on the physical growth of Colombian infants at risk of malnutrition. *Child development, 61*, 29–49.

Susser, M., & Levi, B. (1999). Ordeals for the fetal programming hypothesis. *British Medical Journal, 318*, 885–886.

Suter, P. M. (2012). Alcohol: Its role in nutrition and health. In J. W. Erdman, I. A. Macdonald, & S. H. Zeisel (Eds.), *Present knowledge in nutrition* (10th ed., pp. 912–938). Ames: Wiley-Blackwell.

Sutin, A. R., & Terracciano, A. (2013). Perceived weight discrimination and obesity. *PLoS ONE, 87*, e70048. https://doi.org/10.1371/journal.pone.0070048.

Suzuki, H., Morikawa, Y., & Takahashi, H. (2001). Effect of DHA oil supplementation on intelligence and visual acuity in the elderly. *World Review of Nutrition and Dietetics, 88*, 68–71.

Sved, A. F. (1983). Precursor control of the function of monoaminergic neurons. In R. J. Wurtman & J. J. Wurtman (Eds.), *Nutrition and the brain* (Vol. 6, pp. 223–275). New York: Raven Press.

Swerdlow, J. L. (2000). Modern science embraces medicinal plants. In J. L. Swerdlow (Ed.), *Nature's medicine: A chronicle of mankind's search for healing plants through the ages* (pp. 110–157). Washington, DC: National Geographic Society.

Sydenham, E., Dangour, A. D., & Lim, W. S. (2012). Omega 3 fatty acid for the prevention of cognitive decline and dementia. *Cochrane Database of Systematic Reviews, 6*, CD005379. https://doi.org/10.1002/14651858.CD005379.pub3.

Tacconi, M. T., Calzi, F., & Salmona, M. (1997). Brain lipids and diet. In M. Hillbrand & R. T. Spitz (Eds.), *Lipids, health, and Behavior* (pp. 197–226). Washington, DC: American Psychological Association.

Taibi, D. M., Landis, C. A., Petry, H., & Vitiello, M. V. (2007). A systematic review of valerian as a sleep aid: Safe but not effective. *Sleep Medicine Reviews, 11*, 209–230.

Tanaka, M. (2002). Secretory function of the salivary gland in patients with taste disorders or xerostomia: Correlation with zinc deficiency. *Acta-Oto-laryngology, 122*, 134–141.

Teff, K. L., Young, S. N., & Blundell, J. E. (1989). The effect of protein or carbohydrate breakfasts on subsequent plasma amino acid levels, satiety, and nutrient selection in normal males. *Pharmacology, Biochemistry, and Behavior, 34*, 829–837.

Thomasson, H. R. (1995). Gender differences in alcohol metabolism: Physiological responses to ethanol. In M. Galanter (Ed.), *Recent developments in alcoholism* (Vol. 12, pp. 163–179). New York: Plenum Press.

Thorpe, B. (2013). *Northern mythology, comprising the principal popular traditions and superstitions of Scandinavia, northern Germany and the Netherlands* (Vol. 2). London: Forgotten Books. (Original work published 1851).

Tiwari, B. D., Godbole, M. M., Chattopadhyay, N., Mandal, A., & Mithal, A. (1996). Learning disabilities and poor motivation to achieve due to prolonged iodine deficiency. *American Journal of Clinical Nutrition, 63*, 782–786.

Torun, B., & Viteri, F. E. (1988). Protein-energy-malnutrition. In M. E. Shils & V. R. Young (Eds.), *Modern nutrition in health and disease* (pp. 746–773). Philadelphia: Lea & Febiger.

Townsend, J. W., Klein, R. E., Irwin, M. H., Owens, W., Yarbrough, C., & Engle, P. L. (1982). Nutrition and preschool mental development. In D. A. Wagner & H. W. Stevenson (Eds.), *Cross-cultural perspectives on child development*. San Francisco: W.H. Freeman and Co.

Treasure, J. (2004). *A guide to the medical risk assessment for eating disorders*. London: Maudsley Publications.

Tuthill, A., Slawik, H., O'Rahilly, S., & Finer, N. (2006). Psychiatric co-morbidities in patients attending specialist obesity services in the UK. *Quarterly Journal of Medicine, 99*, 317–325.

UNICEF. (1998). *The state of the world's children 1998: Focus on nutrition*. Available at: http://www.unicef.org/sowc98. Accessed 24 Mar 2015.

US Food and Drug Administration. (1994). *Dietary supplement and health education act of 1994*. Available at: http://www.fda.gov/Regulatory Information/Legislation/FederalFoodDrugand CosmeticActFDCAct/Significant Amendments to the FDCA Act/ucm 1 480003.htm. Accessed 30 June 2015

US Food and Drug Administration, Center for Food Safety and Applied Nutrition. (2002). *Consumer advisory: Kava-containing dietary supplements may be associated with severe liver damage*. Available at: http://www.fda.Gov/foodresourcesforyou/consumers/ucm085482.htm. Accessed 25 Mar 2015

Vitale, M. A., Chen, D., & Kanarek, D. B. (2003). Chronic access to a sucrose solution enhances the development of conditioned place preferences for fentanyl and amphetamine in male Long-Evans rats. *Pharmacology, Biochemistry & Behavior, 74*, 529–539.

Vogler, B. K., Pittler, M. H., & Ernst, E. (1999). The efficacy of ginseng. A systematic review of randomized clinical trials. *European Journal of Clinical Pharmacology, 55*, 567–575.

von Poblotzki, M., Rieger-Fackeldey, E., & Schulze, A. (2003). Effects of theophylline on the patterns of spontaneous breathing in preterm infants less than 1000g of birth rate. *Early Human development, 72*, 47–55.

Wachs, T. D. (1995). Relation to mild-to-moderate malnutrition to human development. Correlational studies. *Journal of Nutrition, 125*, 2245S–2254S.

Wachs, T. D. (2009). Models linking nutritional deficiencies to maternal and child mental health. *American Journal of Clinical Nutrition, 89*, 935S–939S.

Wainwright, P. E. (2002). Dietary essential fatty acids and brain function: A developmental perspective on mechanisms. *Proceedings of the Nutrition Society, 61*, 61–69.

Walford, R. L., Mock, D., Verdery, R., & MacCallum, T. (2002). Calorie restriction in Biosphere 2: Alterations in physiologic, hematologic, hormonal, and biochemical parameters in humans restricted for a 2-year period. *Journal of Gerontology: Biological Sciences, 57*, B211–B224.

Walker, W. O., & Johnson, C. O. (2006). Mental retardation: Overview and diagnosis. *Pediatrics in Review, 27*, 204–212.

Wallner, B., & Machatschke, I. H. (2009). The evolution of violence in men: The function of central cholesterol and serotonin. *Progress in Neuro-Psychopharmacology & Biological Psychiatry, 33*, 391–397.

Wang, Y., & Beydoun, M. A. (2007). The obesity epidemic in the United States – Gender, age, socioeconomic racial/ethnic, and geographic characteristics: A systematic review and meta-regression analysis. *Epidemiologic Reviews, 29*, 6–28.

Wannamethee, S. G., & Shaper, A. G. (2003). Alcohol, body weight, and weight gain in middle-aged men. *American Journal of Clinical Nutrition, 77*, 1312–1317.

Wasink, B., Painter, J. E., & Lee, Y.-K. (2006). Proximity's influence on estimated and actual candy consumption. *International Journal of Obesity, 30*, 871–875.

Wass, T. S., Simmons, R. W., Thomas, J. D., & Riley, E. P. (2002). Timing accuracy and variability in children with prenatal exposure to alcohol. *Alcoholism: Clinical and Experimental Research, 26*, 1887–1896.

Wassenaar, D., le Grange, D., Winship, J., & Lachenicht, L. (2000). The prevalence of eating disorder pathology in a cross-ethnic population of female students in South Africa. *European Eating Disorders Review, 8*, 225–236.

Watson, T. L., & Andersen, A. E. (2003). A critical examination of the amenorrhea and weight criteria for diagnosing anorexia nervosa. *Acta Psychiatrica Scandinavia, 108*, 175–182.

Wecker, L. (1990). Choline utilization by central cholinergic neurons. In R. J. Wurtman & J. J. Wurtman (Eds.), *Nutrition and the brain* (Vol. 8, pp. 147–162). New York: Raven Press.

West, R., Beeri, M. S., Schmeidler, J., Hannigan, C. M., & Angelo, G. (2008). Better memory functioning associated with higher total and low-density lipoprotein cholesterol levels in very elderly subjects without the apolipoprotein e4 allele. *The American Journal of Geriatric Psychiatry, 16*, 781–785.

Westover, A. N., & Marangell, L. B. (2002). A cross-national relationship between sugar consumption and major depression? *Depression and Anxiety, 16*, 118–120.

Whalley, L. J., Deary, I. J., Starr, J. M., Wahle, K. W., & Rance, K. A. (2008). n-3 Fatty acid erhthrocyte membrane content, APOE varesilon4, and cognitive variation: An observational follow-up study in late adulthood. *American Journal of Clinical Nutrition, 867*, 449–454.

Widdowson, E. M. (1985). Responses to deficits of dietary energy. In K. Blaxter & J. C. Waterlow (Eds.), *Nutritional adaptation in man* (pp. 97–104). London: John Libby.

Willner, P., Benton, D., Brown, E., Cheeta, S., Davies, G., Morgan, J., & Morgan, M. (1998). "Depression" increases "craving" for sweet rewards in animal and human models of depression and craving. *Psychopharmacology, 136*, 272–283.

Wilson, G. T., Fairburn, C. G., Agras, W. S., Walsh, B. T., & Kraemer, H. (2002). Cognitive-behavior therapy for bulimia nervosa. *Journal of Consulting and Clinical Psychology, 70*, 267–274.

Wing, R. R., Gorin, A., & Tate, D. F. (2012). Strategies for changing eating and exercise behavior to promote weight loss and maintenance. In J. W. Erdman, I. A. McDonald, & S. H. Zeisel

(Eds.), *Present knowledge in nutrition* (10th ed., pp. 1057–1070). Ames: International Life Sciences Institute.

Winick, M., & Rosso, P. (1969). Head circumference and cellular growth of the brain in normal and marasmic children. *Journal of Pediatrics, 74*, 774–778.

Wolraich, M. L., Lindgren, S. D., Stumbo, P. J., Stegink, L. D., Appelbaum, M. I., & Kiritsy, M. C. (1994). Effects of diets high in sucrose or aspartame on behavior and cognitive performance of children. *New England Journal of Medicine, 330*, 301–307.

Woolsey, M. M. (2002). *Eating disorders: A clinical guide to counseling and treatment.* Chicago: American Dietetic Association.

Worobey, J. (2002). Interpersonal versus intrafamilial predictors of maladaptive eating attitudes in young women. *Social Behavior and Personality, 30*, 424–434.

Worobey, J., & Schoenfeld, D. (1999). Eating disordered behavior in dietetics students and students in other majors. *Journal of the American Dietetic Association, 99*, 1100–1102.

Wurtman, R. J., & Wurtman, J. J. (1995). Brain serotonin, carbohydrate craving, obesity and depression. *Obesity Research, 3*, 477S–480S.

Wurtman, R. J., Hirsch, M. J., & Growdon, J. H. (1977). Lecithin consumption raises serum free choline levels. *Lancet, 2*, 68–69.

Wurtman, R. J., Hefti, F., & Melamed, E. (1980). Precursor control of neurotransmitter synthesis. *Pharmacological Reviews, 32*, 315–335.

Yokogoshi, H., & Wurtman, R. J. (1986). Metal composition and plasma amino acid ratios: Effects of various proteins on carbohydrates, and of various protein concentrations. *Metabolism, 35*, 837–842.

Young, S. N. (1991). Some effects of dietary components (amino acids, carbohydrate, folic acid) on brain serotonin synthesis, mood, and behavior. *Canadian Journal of Pharmacology, 69*, 893–903.

Young, S. N. (2005). Amino acids, brain metabolism, mood and behavior. In H. R. Lieberman, R. B. Kanarek, & C. Prasad (Eds.), *Nutritional neuroscience* (pp. 131–146). Boca Raton: Taylor and Francis.

Young, S. N., Smith, S. E., Pihl, R. O., & Ervin, F. R. (1985). Tryptophan depletion causes a rapid lowering of mood in normal males. *Psychopharmacology, 87*, 173–177.

Yurko-Mauro, K., McCarthy, D., Rom, D., Nelson, E. B., & Ryan, A. S. (2010). Beneficial effects of docosahexaenoic acid on cognition in age-related cognitive decline. *Alzheimer's & Dementia, 6*, 456–464.

Yuwiler, A., Oldendorf, W. H., Geller, E., & Braun, L. (1977). Effect of albumin binding and amino acid competition on tryptophan uptake into brain. *Journal of Neurochemistry, 28*, 1015–1023.

Zeigler, D. W., Wang, C. C., Yoast, R. A., Dickinson, B. D., & McCaffree, A. (2005). The neurocognitive effects of alcohol on adolescents and college students. *Preventive Medicine, 40*, 23–32.

Zhang, J. (2011). Epidemiological link between low cholesterol and suicidality: A puzzle never finished. *National Neuroscience, 14*, 268–287.

Printed by Printforce, the Netherlands